Evolution and Behavior

The relationship between how we evolved and how we behave is a controversial and fascinating field of study. From how we choose a mate to how we socialize with other people, the evolutionary process has an enduring legacy on the way we view the world. *Evolution and Behavior* provides students with a thorough and accessible introduction to this growing discipline.

Placing evolutionary psychology in context with the core areas of psychology – developmental, cognitive and social – the book explores some of the most fundamental questions we can ask about ourselves. Taking students through the principles of natural selection, it provides a nuanced understanding of key topics, such as

- cognitive development and the role of intelligence;
- memory, emotions and perception;
- mental health and abnormal psychology;
- sexual reproduction and family relationships;
- the development of culture.

Addressing a number of controversial debates in the field, each chapter also includes concept boxes, the definitions of key terms, chapter summaries and further reading. This is the ideal introductory textbook for anyone interested in evolutionary psychology. It will not only provide an essential overview of this emerging field but also deepen readers' appreciation of the core tenets of psychology as a whole.

Lance Workman is Visiting Professor of psychology at the University of South Wales in the UK.

Will Reader is Senior Lecturer in psychology at Sheffield Hallam University in the UK.

Foundations of Psychology Series

This series provides pre-undergraduate and first year undergraduates with appealing and useful books that will enable the student to expand their knowledge of key areas in psychology. The books go beyond the detail and discussion provided by general introductory books but will still be accessible and appropriate for this level. This series will bridge the gap between the all-encompassing general textbook and the currently available advanced topic-specific books which might be inaccessible to students who are studying such topics for the first time. Each book has a contemporary focus and fits into one of three main categories including Themes and Perspectives (such as Theoretical Approaches or Ethics), Specific Topics (such as Memory or Relationships) and Applied Areas (such as Psychology and Crime).

Series editors

Philip Banyard is a Reader in Psychology at Nottingham Trent University.

Cara Flanagan is an experienced academic author, writer and freelance author and freelance lecturer in Psychology.

Published titles

Cultural Issues in Psychology
Andrew Stevenson

Ethical Issues in Psychology
Philip Banyard and Cara Flanagan

Essentials of Sensation and Perception
George Mather

Interpersonal Relationships
Diana Jackson-Dwyer

Evolution and Behavior
Lance Workman and Will Reader

Evolution and Behavior

Lance Workman and Will Reader

Routledge
Taylor & Francis Group

LONDON AND NEW YORK

First published 2016
by Routledge
27 Church Road, Hove, East Sussex BN3 2FA

and by Routledge
711 Third Avenue, New York, NY 10017

Routledge is an imprint of the Taylor & Francis Group, an informa business

Every effort has been made to contact the copyright holders for all third party materials used in this book. Please advise the publisher of any errors or omissions, if you are a copyright holder.

British Library Cataloguing in Publication Data
A catalogue record for this book is available from the British Library

Library of Congress Cataloging-in-Publication Data
Workman, Lance, author.
  Evolution and behavior / Lance Workman and Will Reader.
    pages cm
  Includes bibliographical references and index.
  1. Evolutionary psychology. 2. Human evolution. 3. Behavior evolution. I. Reader, Will, author. II. Title.
  BF698.95.W667 2016
  155.7—dc23
  2015014053

ISBN: 978-0-415-52199-4 (hbk)
ISBN: 978-0-415-52202-1 (pbk)
ISBN: 978-1-315-72640-3 (ebk)

Typeset in Arial
by Apex CoVantage, LLC

Contents

List of illustrations

Figures

Tables

Why the fuss?
Evolution and
human behavior

1

What this chapter will teach you

- Evolutionary psychology and proximate and ultimate explanations.
- How evolution by natural selection works.
- Considering evolution from the gene's point of view.
- Sexual selection: How mate choice affects evolution.

Evolutionary psychology is different: Proximate and ultimate explanations

Psychology is the science of human behavior, and as such it has proposed many useful theories as to why humans behave the way they do. One current theory of the psychological disorder **schizophrenia,** for example, is that it is the result of an excess or heightened sensitivity to the **neurotransmitter** dopamine (Kring, Johnson, Davison & Neale, 2014; see Chapter 8); explanations for prejudice center around **in-group–out-group bias** distinctions (see Chapter 5); whereas repeated failed relationships have been explained by the development of different attachment styles in infancy (Chisholm, 1999; see Chapter 6). Although all of these explanations are very different from one another (the first is biological, the second social, the third

KEY TERMS

Proximate explanation In psychology and biology, an explanation for a behavior couched in terms of current biological, cognitive, social or developmental mechanisms.

Ultimate explanation An explanation that attempts to explain why a particular behavior evolved in the first place.

developmental), they are all similar in that they are **proximate explanations.** They take a particular phenomenon and ask what current or past factors led to that particular behavior. In contrast, evolutionary psychology asks for **ultimate explanations** to questions: it asks what are the particular evolutionary processes that led to a person exhibiting that particular behavior?

To illustrate the difference, consider the emotion of disgust. This seems to be a **cultural universal** (it occurs in all cultures that have been studied; Brown, 1991), and there have been quite a few attempts to study it. Some researchers have studied the kinds of things that promote the disgust response, others have studied the neurological causes of disgust and still others have investigated the way that disgust develops through childhood (Haidt, McCauley & Rozin, 1994). But an evolutionary psychologist would ask a different question: what is disgust *for*? Or, put another way, what is the evolutionary benefit to the individual of having a disgust response (we will discuss who the beneficiaries of a particular behavior are later in this chapter).

One answer to this question is that disgust is a solution to a problem called the "omnivores dilemma." Unlike many other species (cows and giant pandas spring to mind), humans are capable of eating a range of very different foods: meat, fruit and vegetables of course, but also grubs, insects, eyeballs and intestines – all of which are perfectly nutritious and, in some cultures, considered a delicacy. But amidst all of the good things to eat are things that are inedible (hair, grass), poisonous (the livers of certain animals, many plants and fungi) or, more importantly, contain pathogens such as bacteria (rotten meat, feces). So how do omnivores such as humans solve this problem? One solution is to consume anything in infancy. Toddlers regularly eat worms, soil and, in some cases, the contents of their own diapers, perhaps relying on their parents to prevent them from eating anything too dangerous. But from about 3 or 4 years of age, a child's dietary repertoire becomes increasingly fixed, favoring things that they are used to and avoiding novel foods (referred to as **neophobia**), and many novel foods are seen as disgusting. So, disgust prevents contamination from potentially dangerous materials. It is an extremely aversive emotion, first encouraging individuals to withdraw from the object, and – should it enter the mouth – promoting the gag response, the purpose of which is to expel the offending item from the body. For those readers who are unclear as to what this process involves, watch an episode of *I'm a Celebrity . . . Get Me Out of Here!*

Disgust has, of course, subsequently become expanded to take on a moral dimension (think of incest or child abuse).

DISCUSS AND DEBATE – NEOPHOBIA

Think of times when you have been exposed to new foods and what you initially made of them. Did you immediately like them? Did you consider them repulsive? Did you eventually overcome your revulsion? If so, how, and why? Peer pressure and the desire to impress other people is often a motivation (say, to impress your peers or a member of the opposite sex). But neophobia can have remarkably deep roots – something that makes entertaining television when junk-food junkies are asked to swap to a healthy diet.

So we can see here a potential answer to the ultimate question as to what disgust is *for*. It is an evolved mechanism to prevent us from putting the wrong kind of thing in our mouths, which, as omnivores, is a potentially very large set of things indeed. As an aside, we might apply this line of thinking to pleasant and unpleasant sensations. Why do some things feel, taste, smell, look or sound good and some things the opposite? There is nothing about feces which makes it intrinsically bad smelling; rather, the aversive smell is the result of the above adaptation: it smells bad so that we avoid it. Doubtless to a dung beetle, nothing else smells so sweet.

DISCUSS AND DEBATE – PROXIMATE AND ULTIMATE EXPLANATIONS

As an exercise, try thinking of some psychological topics and consider proximate and ultimate questions that might be asked of them. To get you started, here are a few examples: out-group prejudice, laughter, schizophrenia and forgetting.

Where ultimate explanations come from: Evolution by natural selection

Ultimate explanations only make sense because evolution has a well-worked-out theory: **natural selection.** Before discussing this, it is useful to spend a few words discussing what a theory actually is.

The purpose of a theory is to explain things in the world, and some theories – and the theory of evolution by natural selection is certainly one of them – are supported by a great deal of evidence. So, if you are thinking that theory necessarily entails some kind of vague guess, it does not. As we shall see, evolution by natural selection is precise and is supported by mountains of scientific evidence. So much evidence, in fact, that were it to be proved to be untrue, scientists would be almost as surprised as if they were to find out that sun is not the center of the solar system.

Natural selection was Darwin's theory. Although Darwin's contemporary, Alfred Russell Wallace, had a very similar idea, it was Darwin who managed to get the theory published first in 1859's *On the Origin of Species*. During his famous circumnavigation of the world on board the *HMS Beagle* between 1831 and 1836, Darwin noticed both the great diversity of life and the fact that members of each species seemed to fit their environments so well. It was as if each organism had been designed to deal with the challenges of its particular environment. Figure 1.1 shows what have come to be known as *Darwin's finches* that live on the Galápagos Islands, famously visited by Darwin

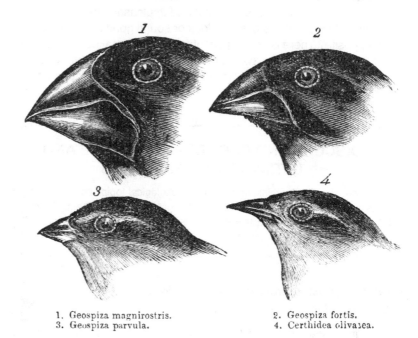

1. Geospiza magnirostris.
2. Geospiza fortis.
3. Geospiza parvula.
4. Certhidea olivacea.

Figure 1.1 Darwin's finches, by John Gould.
Source: John Gould [Public domain], via Wikimedia Commons.

during his voyage. The birds' bills are well designed to exploit their particular **ecological niche** (and in particular to solve the foraging challenges of each Island). Before you read on, can you guess what each of the four birds eat?

The answers are that (1) primarily eats nuts, (2) eats grain and (3) and (4) both eat insects. And the bill design reflects this, depending on whether they need a nutcracker, a grain crusher or a more delicate instrument for persuading insects out of their holes. As well as partly inspiring Darwin to produce his theory of evolution, changes in their physical structure as the habitat of each Island has changed since Darwin's death has provided evidence for evolution in action.

Darwin's theory

On his return to England, Darwin spent many years formulating how this fit with the environment had come about. Many people prior to Darwin had attempted to solve the same puzzle and produced various solutions. The philosopher William Paley concluded that complex design, such as that found in living organisms, could only be explained by their being produced by divine creator (Paley, 1802). Although notionally a Christian, Darwin rejected this approach as unscientific: science should explain phenomena in terms of natural rather than supernatural processes. Darwin's task was to explain how design could come about without there being a designer.

He called the process that he discovered *natural selection*. The name comes by analogy with artificial selection, which is where plants or animals can be changed over time by humans selectively cross-breeding them for desired **traits** (see Figure 1.2). Natural selection involves no designer, and it is as stunning in its simplicity as it is far-reaching in its implications. It hangs on three principles.

The first is **heritability**. Offspring tend to resemble their parents more than they do other members of the species. Something is clearly passed from parent to offspring that causes this similarity. Now we call this something **genes,** but Darwin knew nothing about genetics at the time.

The second is **variation**. Although parents tend to resemble their offspring, they differ in subtle but sometimes important ways. There are a number of sources of variation. Mutations are the simplest. We now know that genes play crucial role in determining an individual's make-up, and we know that genes are made of **deoxyribonucleic acid (DNA;** see Chapter 2). DNA is a large and very complex chemical, and

Figure 1.2 Sweet corn has been artificially selected for its sweetness, size and color. The cob on the left is the original wild version, the one in the middle is after a degree of cross-breeding and the one on the right is the current form.

Source: John Doebley [CC BY 2.5 (http://creativecommons.org/licenses/by/2.5)], via Wikimedia Commons.

as such its structure can be changed by things such as radiation or the presence of other chemicals. Nuclear accidents such as that which occurred at the Chernobyl nuclear power station in 1986 have led to an increase in genetic abnormalities and birth defects as a result of mutations caused by radiation. But, radiation is present at low concentrations everywhere, all the time, which can cause changes in the genetic material of offspring, causing them to be subtly different from their parents.

In addition to mutations, another source of variation is sex. Bacteria make copies of themselves, but humans and many other organisms do not. Instead they combine the genes from two individuals to produce a unique combination in the offspring (see Chapter 2).

The third principle is **differential reproductive success,** which is a bit of a mouthful, but is in many ways Darwin's most important insight. As we have seen, offspring are different from their parents, and this can have an effect on the organism's ability to survive and reproduce. In many (probably most) cases these differences have either no effect or – as we see with birth abnormalities – reduce survival and reproduction chances, but in some cases the offspring can be better at surviving and/or reproducing than their parents are. And, of course, they will tend to pass these on to their offspring (heritability again), and in the long run the new form may well replace the old form.

The textbook example of this is the peppered moth (*Biston betularia*). As its name implies, this moth traditionally has a "peppered" or speckled coloration, all the better for hiding on the lichen that clings to the bark of the trees in the forests, which are the moth's main habitat. Lichens are plant-like organisms that are, in fact, a partnership between a fungus and a photosynthetic organism such as an alga (plural, *algae*). They are very sensitive to air pollution, which means that they only grow in profusion where the air is relatively clean. As a result of industrialization in Britain through the 19th and 20th centuries, air pollution killed off much of the lichen on trees in and around cities, revealing the brown bark underneath. Suddenly, the peppered moth, so well camouflaged on the lichen-covered trees, was now conspicuous to predators, such as birds, and could have been driven to extinction. Fortunately, a new variant emerged, which was dark brown in color (see Figure 1.3), making it blend in more effectively than the peppered

Figure 1.3 The peppered moth in speckled (peppered) and dark forms.

variant and therefore less likely to be discovered by predators. This variant, being more effective at survival, was more likely to pass its genes on to the next generation (differential reproductive success), which would resemble the parent (heritability) and its numbers grew as the peppered form declined.

It must be stressed that the change in the environment in no way produced the mutation for dark coloration, all it did was "select for" the dark coloration gene over the speckled by virtue of dark forms being better camouflaged and thus more resistant to predation, and therefore more likely to reproduce and pass on the genes for dark coloration. Doubtless that prior to industrialization, dark-colored mutations arose, but on lichen-covered trees these forms were more conspicuous and therefore less successful than their speckled counterparts.

We can therefore see that, counter to what some people might say, the process of natural selection is not random. Mutations and other sources of variation might well arise by chance, but the preferential selection of some forms over others is dependent upon the extent to which they fit their environment. This second part is not random, therefore.

You might at this point be starting to doubt our claim that natural selection is a very simple process, but the aforementioned description is the theory in its entirety – barring a few minor but important details, which we cover in subsequent chapters. The evolutionist Richard Dawkins (2008) manages to summarize the process in a single sentence.

> Given sufficient time, the non-random survival of hereditary entities (which occasionally miscopy) will generate complexity, diversity, beauty and an illusion of design so persuasive that it is almost impossible to distinguish from deliberate intelligent design.
> (Dawkins, 2008, http://www.theguardian.com/ science/2008/feb/09/darwin.dawkins1)

Packed into that statement is everything we discussed above: heritability (heredity), variation (miscopying) and differential reproductive success (non-random survival). As we said previously, natural selection is a simple process: summarizing some of the other famous theories of recent times in such a pithy way, such as Einstein's theory of general relativity, Freud's psychodynamic theory or quantum mechanics would be far more challenging.

From small to big changes: The evolutionary process over billions of years

The evolutionary changes experienced by the peppered moth are trivial compared to some of other changes that have occurred during the history of life on earth, such as the evolution of the eye, bipedalism (see Chapter 2), digestion and language, but it is the contention of modern biology that all of these were produced by exactly the same process described above, with the extra ingredient of time. And there has been so much time: current research estimates life to have existed for perhaps as long as 3.5 billion years (the earth is "only" 4.5 billion years old, so life seemed to start pretty quickly). To get some idea of the scale of this, we modify an example from Richard Dawkins.

Imagine that all your ancestors are standing in a line, with you at the head. Standing a meter behind you is your mother, and a meter behind her stands her mother and so on. Now imagine that you turn and walk down the line. As you progress, you reflect upon how fashions have changed over the years, assuming, for ease of discussion, that each mother gave birth to your ancestor when she was 20 years of age. Then, after walking for no more than 5 meters, we have already walked back in time 100 (5 × 20) years, and our female ancestors are wearing long dresses and petticoats. After walking for just over 20 meters, your ancestors will be wearing Tudor-style clothes; after only 150 meters, we are into the Bronze Age; and as meters turn into kilometers, the Stone Age. We have now walked 10 kilometers, and although fashions have degraded rapidly, from simple cloth to animal skins and, possibly, to no clothes at all, all our ancestors are recognizably human. But, as we progress, we notice that gradually, as kilometers go on, they start to change, becoming slightly more apelike – hairier and less upright in posture. Rather like the movement of the hour hand on a clock, you would only notice change in retrospect; you would be unaware of any changes from moment to moment (or from kilometer to kilometer). It is worth repeatedly stressing how slow evolution is on the human timescale.

Somewhere around the 250 kilometer mark (give or take a few tens of kilometers), we come across an important person. This person is not just your ancestor but the ancestor of all humans – so when someone tells you that they are related to someone famous, tell them that you are too – we all are, it's just that you don't get sent a Christmas card. Standing immediately behind her is her mother and someone possibly even more prestigious. We know that this woman had at least two daughters. One of these you have already met – the ancestor of all humans – the other, her sister, was the ancestor of all chimpanzees. The other sister isn't here because she isn't your ancestor, but even so, she would look no different from your ancestor – or at least no more different than any sisters normally look – she would not be any hairier or have any more of a penchant for bananas. The technical term for an individual who is the ancestor of two or more lineages is the **common ancestor**.

> **KEY TERM**
>
> **Common ancestor** An individual that gave rise to two or more separate lineages. For example, the common ancestor of chimps and humans, the common ancestor of plants and fungi (and animals, too – fungi are more closely related to humans than they are to plants!)

It is worth pausing for a second to reflect upon the simple fact that a female gave birth to two sisters (they could even have been identical twins), each of whom ultimately gave birth to two separate lineages. How do we know this? We, as far as we know, have not dug up these common ancestors. Fossils are the exception rather than the rule, and we are lucky that we have as many as we do. No, the reason why we know this is simply logic. If there was a common ancestor of chimps, then it *simply must be the case* that one of her children founded the chimp dynasty and the other the human dynasty because this is what the term *common ancestor* means.

You could, if you like, take a diversion and follow the chimpanzee lineage through your ancestor's sister, this time walking in the opposite direction forward in time; her descendants would, for quite a large distance, look no different from the descendants of your ancestors. Gradually, as the kilometers wore on, they would start looking more like chimps: natural selection does not happen overnight. We do not see animals giving birth to offspring that differ dramatically from their parents (frogs giving birth to mice, for example). It happens slowly, by tiny changes that step-by-step mount up and lead to dramatic changes. Incidentally, you will see nothing that resembles, say, a gorilla, on this diversion down the chimp lineage. Despite external appearances, chimps are more closely related to humans than they are to gorillas (see Chapter 2).

DISCUSS AND DEBATE – ANIMAL ETHICS AND GENETIC RELATEDNESS

The genetic closeness between humans and our closest living relatives, the chimpanzees, has led to some to argue that we should extend human rights to include chimpanzees. What do you think of this? Does a system of ethics based on relatedness make sense, or should we rely on more traditional means, such as capacity to suffer or intelligence.

If we abandon the detour and return to the ancestor of all humans we could continue on our backward journey, and after an extremely long walk of around 20,000 kilometers, which is a little more than the distance from the UK to New Zealand, we would meet the common ancestor that we share with fish. But that's a very long way, and you've probably got the point now. Two points, in fact, worth reiterating. First, although small changes (e.g., in coloration, such as the peppered moth) can occur in a few years or decades, bigger changes (bipedalism, wings, brains) take much longer. However, there has been a very large amount of time and a very large number of individuals for small changes to add up to big changes. The second point is that we really are all related to each other, and if we traveled far enough back to around 3.5 billion years ago, we could meet the common ancestor of every living thing – animals, plants, bacteria, the lot. An organism referred to as the **last universal common ancestor**. Attempting to put this into kilometers traveled is difficult as, for our purposes, we were assuming parents gave birth to offspring at 20 years of age. While this is a reasonable assumption for primates such as humans, we know that many other organisms reproduce much more rapidly, in just a few years, months or, in the case of bacteria, sometimes minutes. For this reason, even our estimate of 20,000 kilometers for the common ancestor for fish is a rough estimate.

KEY CONCEPTS BOX 1.1

Some evidence for evolution and the commonality of all living things

How do we know that everything is related? There are a number of lines of evidence. First, and probably most obviously, there is the fossil record. There are fossils of animals that, for example, show evidence of characteristics found in both fish and land animals, which suggest that they might be a common ancestor. Even in the absence of possible common ancestors there are marked similarities that can be found between animals (and plants) that are superficially very different. All vertebrates, for example, show evidence of four limbs, even in animals such as snakes, where limbs have been lost. Embryological research is of help here too, as animals that might be very different when adults can be markedly similar when they are developing. Looking at the embryos of fish, turtles, salamanders and humans, it can be seen that although they end up looking very different, they start very similar – early in **embryogenesis,** humans even have tails!

KEY TERM

Embryogenesis is the process of development of the embryo (*genesis* means "origins").

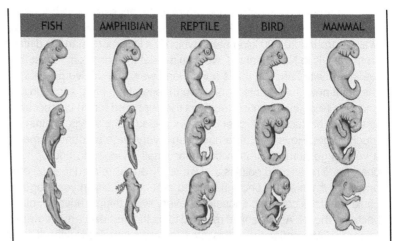

Figure 1.4 The embryos of various vertebrates at different ages.
Source: Image copyright © stihii/Shutterstock.com.

A third source of evidence is genetic. As we examine the genetic machinery of a variety of species it shows how similar we all are. For example, genes that control some of the basic physiological processes, such as "Krebs cycle" (concerned with respiration, which is the process of producing energy from food) are the same across pretty much all species (mammals, reptiles, plants, fungi, bacteria), suggesting that we inherited such genes from ancestors billions of years ago.

Darwin understood the commonality of life, although he had no real evidence to support it, as he said:

> Therefore I should infer from analogy that probably all the organic beings which have ever lived on this earth have descended from some one primordial form, into which life was first breathed.
>
> (Darwin, 1859, p. 140)

The "selfish gene"

One of the more complicated but exciting findings of evolutionary thinking is the so called **gene-centered** view of life, or the selfish gene. As we shall see later, taking the perspective of the gene itself clears up many mysteries. Before we explain this, however, it will help to consider what life is and how it began.

In the beginning . . .

Defining life is very difficult. Biology text books tend to fall back on the seven characteristics of living things (respiration, reproduction, loco-motion, made of cells, react to the environment, excretion and growth), but this defines life as it is now, not necessarily life as it began, which is what we are interested in. Of these seven characteristics, reproduc-tion is the most important, because if no living thing ever reproduced, evolution could simply not happen, and life would not have got off the ground (recall that making copies is one of the foundations of Darwin's theory).

We don't know what the first form of life was, but one thing that we do know about the origins of life is that it began as a simple **replicator**. A *replicator* is a word that is used for something that makes copies of itself; for example, cells and computer viruses are replicators. It is assumed that life arose from relatively simple chemicals that had this property. Figure 1.5 shows a schematic diagram of how a simple replicator might work.

We imagine some kind of chemical soup in which there are a number of different, sep-arate molecules (labeled, A, B and C; see Figure 1.5a). These chemicals do not nor-mally react and join together, but somehow, due to extreme temperatures, pressure or the presence of some **catalyst**, the one set of chemicals has formed itself into a chain (see Figure 1.5b). This, then, itself serves to catalyze further reactions. Free-floating A chemicals are drawn to the A chemical on the chain and likewise for Bs and Cs (see Figure 1.5c). Catalyzed by the chain molecule, the previously separate A, B and C form bonds with each other (see Figure 1.5d), and then float off, pre-sumably to repeat the process (see Figure 1.5e). Although rare, there is nothing particularly magical about this process, and, at root, the overall process is pretty much the same for DNA replication (see Chap-ter 2). What is rather stunning to consider, however, is that something like what is depicted in Figure 1.5 – just chemicals in a watery environ-ment, ultimately gave rise to the person who is reading this book.

As before, the timescales are barely imaginable, but gradually the replicating chemicals doubtless became more complex (possibly the result of copying errors, such as those described above) until they

Figure 1.5a

A-B-C

Figure 1.5b

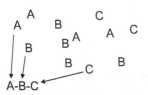

A-B-C

Figure 1.5c

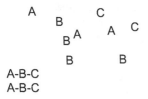

A-B-C
A-B-C

Figure 1.5d

A-B-C
A-B-C

Figure 1.5e

Figure 1.5 Mutation and natural selection diagram.

resembled something like modern DNA. Then, another boundary was crossed – the presence of other chemicals led to the replicator being able to create a boundary between itself and the outside word – the first cell wall. This permitted all sorts of other chemical process to occur without their potentially useful products floating away into the surrounding liquid environment. Finally, after many millions of years, we have something that would be recognizable to us as the first living cell.

The gene-centered view of life: Replicators and vehicles

We are so used to thinking about life at the large-scale level – mammals, trees and birds – that we often fail to appreciate that fundamentally life is just a series of chemical processes, but these processes – such as replication, energy generation (e.g. through respiration) – all had to be established before things we could recognize as life happened. For this reason it is not surprising that we share many of our genes with simple organisms such as bacteria (see above).

Through the process of natural selection described above, these early cells increased in complexity and eventually evolved the useful knack of hanging together in groups rather than going their separate ways – **multicellularity.** This was probably the last of the biological "big bangs," and once multicellularity had evolved, it opened the door for the eventual evolution of the animals, plants and fungi that we see around us today.

But now consider another thing. You probably don't have the time or inclination to think about your genes very much and what their function is. If pressed, however, you might say something like "they produce useful chemicals for my body" or "they played a role in putting my body together." In other words, we tend to think about genes as doing useful things for us. The description above suggests that, on the contrary, we serve their interests rather than the other way round. Think of it: if genes managed to find a profitable niche by building cell walls, could we not also think of them finding a similarly productive niche by producing organisms that had language and consciousness?

This idea is known as the gene-centered view of life and sees organisms as "vehicles" (Dawkins, 1976/2006) assembled through natural selection by genes for their own ends, which is ultimately passing on more copies of themselves. Genes, of course, do not have motives, the process is the same as that for the peppered moth. Genes that put together bodies with brains that are able to produce language seem to have found a new niche to exploit, possibly out-competing other languageless primates (see Chapter 6).

Hamilton's rule

One of the key figures in the development of the gene-centered view of life was biologist William Hamilton (1964). What puzzled him was why animals seem to cooperate. Famously, social insects such as ants, bees and termites cooperate so completely that one will lay down its life for its nestmates. Something that seems incoherent if we think – as many did – of organisms as being selfish and only out for themselves. What Hamilton showed was that cooperation can occur if we focus on the gene rather than the individual organism.

Consider childcare in humans. This is costly to parents, so why do they bother? You might respond by saying it is because they love their children, but we are still faced with the problem as to why parental love evolved – why would you design an animal to do such a thing? It all makes sense from the gene's perspective. Imagine that originally humans didn't care for their children. Once they had produced offspring they looked after them, but without much enthusiasm. The parents would do it grudgingly and half-heartedly. In such a situation, many babies would probably die through neglect. Now, imagine that by some mutation a mother thought her baby was the most beautiful thing she had ever seen. She wanted to be with it all the time and do everything she possibly could for it. The increased parental care received by this child would almost certainly put it at an advantage compared to the other, neglected children. Crucially, this baby stands a 50% chance of inheriting the childcare gene from his mother (males and females contribute half of their genes, so any gene that is present in one parent but not the other stands a 50% chance of being in the child – see Chapter 2). The increased survival chances of this child compared to the others mean that in all likelihood the childcare gene – just like the gene for dark coloration in the peppered moth – will start to proliferate.

A similar mechanism probably led to altruistic behavior in ants, wasps and bees, where all members of a hive or nest are close genetic relatives. Hamilton showed that in close relatives self-sacrificial behavior can evolve. The self-sacrificial behavior of the male red-backed widow spider can be explained in a similar way. Genes are influencing the behavior of one organism in order to help copies of themselves in another. This notion has led to the concept of **inclusive fitness** (see, also, Key concepts box 1.2). Fitness is a measure of how many copies of itself a gene passes on to the next generation. An organism can pass copies of genes on directly through reproduction or indirectly by helping close relatives to reproduce (e.g., the likelihood that a full sibling has one of your genes is on average 50% – the same likelihood for parent and offspring). Thus inclusive fitness is inclusive because it includes those genes passed on directly and those passed on indirectly through helping genetic relatives. We examine this point more closely in Chapter 2.

Sexual selection

Sexual selection is sometimes referred to as Darwin's "other theory." In 1871, Darwin published *The Descent of Man and Selection in Relation to Sex.* This was an attempt to fill in some of the holes that seemingly couldn't be explained by his original theory, natural selection. Some things simply did not make sense. The peacock's tail is the classic example. This is an expensive item in a number of ways. First, it takes energy to grow such a large and ornately patterned tail, and energy needs to be obtained from food, which requires foraging. Second, it needs to be maintained, which again takes time and effort. Third, it is a substantial encumbrance to the bird, making it harder to get around and flee from predators, so there is an additional predation risk. All of these costs suggest that it must serve an important function, but if it does has a function it doesn't seem to be concerned with survival. On the contrary, it may well decrease the animal's chances of survival for reasons already discussed. The answer boils down to sex.

Above, we saw how natural selection provides design without a designer, nobody is choosing one **phenotype** over another and some variants win out simply because they are better fitted to the environment, as for the tale of the peppered moth. But once sex evolved, this was no longer the case. As we saw above, sexual behavior placed a **selection pressure** on individuals to choose with whom they have offspring. This means that, in a very real sense, the evolution of sex brought with it the presence of actual choosers. When we ask "Who made the peppered moth dark?" the answer is no one;

KEY TERMS

Genotype and **phenotype** *Genotype* is the word used to describe the genetic make up of an individual (the set of genes that an individual carries); *phenotype* is used to describe the observable characteristics and traits of that individual, including physical and behavioral traits. In some cases an individual can carry genes that are not "expressed" (do not appear) in the phenotype.

Selection pressure (or sometimes **selective pressure**) A force that seems to be acting upon organisms, driving them to evolve in a particular way. For example, darker trees place a pressure on peppered moths, forcing them to evolve into a darker form. This is an admittedly useful metaphor and should not be taken literally. The "force" is nothing more than some genes (e.g., genes for dark coloration) being more successful in a particular environment than other genes (e.g., for speckled coloration). It is another example of the non-random survival of hereditary entities.

as we saw above, it is simply the differential reproduction of certain phenotypes, but if we were to ask "Who made the peacock's tail large and colorful?" the answer is peahens: the females of the species. Sexual selection has left its mark in other ways. As well as males advertising their fitness through costly ornamentation, as for the peacock's tail, some males will compete with one another for access to females, which has given rise to adaptations for intermale conflict, such as horns or antlers, large physical size and very possibly behavioral adaptations, such as specialized neural circuitry for intermale aggression (see Chapters 3 and 4).

So why do animals choose and compete to be chosen? To answer this question we need to start with the realization that not all genes are equal. Some produce traits that are well fitted to the environment in which the organism lives. For example, the genes for dark coloration in the peppered moth are wonderfully designed for lichen-less trees; the genes for speckled coloration, less well designed for that environment. Using Darwin's terminology we can describe the first set of genes as more fit or of a higher fitness to the environment than the latter. Genes can vary in fitness in other ways. Some animals' immune systems are stronger than others due to genetic factors – some animals are stronger, some more intelligent and so on.

Imagine, for the moment, that animals were indiscriminate with regard to who they mated with – they did it willy-nilly. Some animals would mate with healthy animals, which would generally produce healthy offspring, and others would mate with unhealthy animals, which would tend to produce less healthy offspring. Now imagine that somehow, through some mutation maybe, an animal managed to discern how good a potential mate's genes were. It could not do this directly, but it could do it indirectly, for example, by examining cues relating to general health. Are its eyes clear and bright? Is its skin clear and free from signs of disease, which might be indicative of a compromised immune system? By choosing sexual partners based on indicators of health and general fitness, the offspring of such animals would be more likely to thrive than the offspring of those that could not discriminate. As well as being healthier, these offspring would also likely inherit the ability to choose from the mutant parent, and choosy animals would now start to proliferate at the expense of animals that could not choose. Pretty soon (in evolutionary terms) choice would be the default.

The upshot of what was doubtless a very long evolutionary process seems to be that we have adaptations that are concerned with choosing mates and being chosen. We cover the intricacies of this in Chapters 3 and 4, but suffice to say that many animals – and humans are no exception – spend a considerable amount of time and energy dedicated to making themselves more attractive to the opposite sex. This might include gathering resources, increasing status, preening and engaging in displays of dexterity, strength or competence.

DISCUSS AND DEBATE – WHAT DO MEN AND WOMEN REALLY WANT?

Sexual selection theory suggests that physical and psychological traits are shaped by their being favored by members of the opposite sex. Consider what people generally find desirable in a potential mate (male and female) and how those traits might be indicative of underlying genetic quality.

Offspring survival

Above, we have seen that evolution not only provides adaptations that aid the survival of the organism but also reproduction. But it would a mistake to think natural selection's work is done once an organism has had sex; many organisms have adaptations that try to ensure that offspring themselves survive.

As we saw above, childcare in mammals such as humans can be seen as an adaptation that increases the chances of copies of genes going forward into the next generation. Childcare is so important to mammals that it defines them: the definition of a mammal is that they produce a special food, milk, for their offspring. But childcare is not exclusively mammalian. A bird's nest is an adaptation to ensure that their young are kept safe before they can fend for themselves while their parents feed them; the fish world has species who keep their offspring in their mouths and ants have a special caste – known as nurse ants – whose explicit purpose is the care of the young.

Childcare, such as those above, is only one strategy of ensuring offspring survival and ultimately reproduction. Most other species ensure survival simply by having a vast quantity of offspring (see Chapter 6) – think of baby turtles running to the sea in their hundreds, avoiding birds and other predators. Most will not make it, but on average enough will survive for the lineage to continue. Even in such negligent species, natural selection has ensured that there is rather more to infant survival than sheer weight of numbers. Consider the laburnum tree. Like a turtle, it produces a vast number of offspring, this time in the form of seeds. Natural selection has equipped the parent plant with a mechanism for aiding the survival of these offspring. When the seed pods become ripe they explode, catapulting the seeds away from the parent so that they don't compete with the parent for food and light – a competition that the offspring would be destined to lose.

KEY CONCEPTS BOX 1.2

Survival of the fittest: What this means

Evolution by natural selection is often described by the slogan "survival of the fittest" – which was used by Darwin following the recommendation of his friend the academic Herbert Spencer. Although the term is familiar to many, it is unfortunate in that it is frequently misunderstood. Many imagine that what is surviving here are organisms, and they do so on the basis of winning some evolutionary version of circuit training where only the most pumped up and sculpted individual can survive. But as we have seen, survival is only one piece of the evolutionary jigsaw – reproduction and offspring survival are equally important. What Darwin was referring to here was not the survival of individual organisms but those "hereditary entities," what we now know as genes – and as we saw above some of your genes, those that you share with bacteria, seem to have survived for billions of years. And fitness is nothing to do with bench presses but rather relates to the extent to which the gene produces traits that "fit" the environment (think again of the coloration genes of the peppered moth).

So the individual organism does not evolve, it merely survives and has offspring. What evolves are genes, which change slowly over time, causing small but significant changes to the success (in terms of survival, reproduction and offspring survival) of the organism in which they are embodied. We will return to this notion of the level at which natural selection "selects" in the final chapter.

From evolution to evolutionary psychology

Most people are generally comfortable with the idea that our physical selves evolved through a process of natural selection but are uncomfortable with the notion that behavior evolved. As we saw above with our examples of disgust, childcare and mate choice, there is no reason why behavior should be any different to physiology (we give many more examples in subsequent chapters). At root, evolutionary psychology tries to do two related things. First, it attempts to understand the behavior of humans and other animals by considering them as being designed to fulfil an ecological niche. Ancestral humans repeatedly encountered problems, such as how to find a mate, how to find food, how to avoid premature death, how to ensure their children survived, and any genetic changes (e.g. through mutation) which helped solve these problems would be passed on to their offspring. Related to this is to find "ultimate" explanations of behavior: to try to determine why a particular trait or behavior evolved as well as describing the developmental, cognitive, neuropsychological or social mechanisms (see, again, disgust, above).

We must be clear that this does not deny the existence of social or individual learning, or cultural factors. Evolutionary psychology does not prohibit the existence or importance of learning mechanisms; it just tries to ask the extent to which genetic explanations might also be relevant in understanding psychology. Furthermore, evolutionary psychology can raise important questions about the very nature of learning. Naively, people sometimes ask whether some behavior is the result of genes or learning. If the data show that the behavior is entirely learned, then it is assumed that genes play no role the acquisition of the behavior; but this is inaccurate. Genes may not be directly responsible for that *specific* behavior, but they were certainly responsible for shaping the brain that is capable of learning it. It is worth reflecting on this point a little. The ability to learn, by which we mean to modify behavior based on past experience, has massive benefits in terms of survival and reproduction, without this ability organisms would be doomed to repeat past failures and couldn't benefit from past successes. So, the ability to learn was itself something that evolved: in many animals, such as humans, spectacularly so. In subsequent chapters we also discuss the extent to which learning mechanisms might have evolved that were designed to focus learning on the things that are important to our species (such as disgust, as above, but also language and many other things).

Summary

- Evolutionary psychology concerns itself with ultimate questions (for what purpose did particular behaviors evolve?) as well as proximate ones (what are the more immediate causes of behavior). Ultimate questions are not "better" or more important than proximate questions, they are just questions at a higher level (longer time frame).

- Darwin's theory of evolution by natural selection explains the evolution of complex life from simpler forms. The process has three aspects: heritability, variation and differential reproductive success. Together, these three components can account for the evolution of complex life and complex behavior.
- Darwin speculated on the connectedness of all living things, suggesting that all organisms are descended from a single, common ancestor. Subsequent genetic and developmental evidence confirms this.
- The gene-centered view of life and **Hamilton's rule** show how behaviors can evolve that are not in the interest of the individual organism who produces that behavior. For example, some animals will sacrifice themselves for their offspring or other genetic relatives.
- Sexual selection – Darwin's "other" theory, shows how organism's choosing mates can affect how traits evolve. The peacock's tail has been shaped not to aid the individual's survival but to appeal to females. Female choice has therefore left its mark on many species.
- Evolutionary psychology is not an alternative theory to learning; rather, it can help explain how learning happens and what learning is for. The ability not to repeat past mistakes was a very important evolutionary step.

FURTHER READING

Dawkins, R. (2004). *The Ancestor's Tale: A Pilgrimage to the Dawn of Life.* London: Phoenix.
Dawkins, R. (2006). *The Selfish Gene.* Oxford: Oxford University Press.

How did we get here

Human evolution and genetics

2

What this chapter will teach you

- The "stages" of human evolution.
- Theories of why we evolved such a large brain/high level of intelligence.
- Mendelian and post-Mendelian genetics.
- Introduce the relationship between genes, brain and behavior.
- The notion of heritability.

In this chapter we will consider the pathway that our ancestors took since diverging from our **common ancestor** with the chimpanzee. We will also consider two main types of hypothesis that have been developed to explain human intellectual evolution – that is, ecological and social complexity models. We will then introduce Mendelian genetics and developments in the field since the time of Mendel. Finally we will introduce the concept of **heritability** and begin to examine the relationship between genes and behavior.

Human evolution

KEY CONCEPTS BOX 2.1

Darwin and human evolution

While in the *Origin of Species* (1859) Darwin had devoted just one sentence to human evolution, during the early 1870s he produced two entire books that considered the subject. In 1871 he published *The Descent of Man and Selection in Relation to Sex,* and in 1872 *The Expression of the Emotions in Man and Animals.* Both books made links between ourselves and other species – the former concentrating on fleshing out his notion of sexual selection as a driving force for evolution (see Chapter 1), and the latter being concerned with commonalities of emotional expression between us and other species. Such commonalities suggested to Darwin that we share a common evolutionary ancestor with all other species. Many have speculated as to why Darwin, who had previously appeared reluctant to examine human origins, suddenly decided to pour forth on the subject. It has been suggested that his initial hesitation was related to a desire not to offend the church, but having become the most famous scientist in the world by 1870 he no longer felt constrained by such considerations. The answer is probably more simply that primitive human-like fossils were uncovered or identified during the intervening years.

Five years after the publication of *The Origin of Species* in 1864, the first Neanderthal fossil was identified by Irish geologist William King. King considered this and further similar finds to be a subbranch of human evolution – the *Neanderthal* people (so named due to having been unearthed in the Neander valley of Germany – see below). Following this, in 1868, the remains of the first early modern humans were found at Cro-Magnon, in France, by French geologist Louis Lartet – leading to the label of the *Cro-Magnon* people. The uncovering of such fossils meant that early evolutionists such as Darwin now had at their disposal both the mechanism for change in a species (**natural selection**) and clear evidence that more primitive human-like forms had previously existed (a small number of fossil remains). This, in turn, meant that it was possible to begin to piece together the sequence of changes that made up human physical evolution. Given the paucity of remains at that time, however, Darwin's latter two evolutionary books (see Key concepts box 2.1) did not attempt a detailed sequence of the "stages" of human evolution but instead considered the driving forces that led to our divergence from the great apes. In this way, Darwin was making suggestions as to why humans changed from their simian relatives in evolving such a large brain and the language and culture that it supports. We will consider a modern understanding of potential evolutionary driving forces for such changes after first considering the evolutionary pathway that our ancestors took since their departure from the great apes.

Becoming human – Tracing our ancestry

The early fossils that Darwin had at his disposal have been eclipsed by subsequent finds – with recent discoveries now allowing us to put together quite a detailed picture of human physical evolution since our split from the great apes. Humans, of course, did not evolve from our nearest relative, the chimpanzees, but rather we both shared a common ancestor somewhere between 5 and 6 million years ago (see Chapter 1). The earliest fossil **hominin** (human-like ancestor) uncovered in Ethiopia – *Ardipithecus ramidus* (*Ardipithecus* loosely translates as "ground ape") dates back to 4.4 million years ago. *Ardipithecus* stood about three and a half feet tall and had long, robust, ape-like arms but also relatively small human-like teeth (White et al., 2009). Based on the **phalanges** (foot bones) uncovered, *Ardipithecus* was clearly able to walk upright but also (due to its long, robust arms) was an able tree-climber. In this sense, it was an intermediate between humans and chimps. Given, however, it had a brain capacity in the region of 300 cc in terms of intellectual abilities, *Ardipithecus* was much more similar to chimps (this is virtually identical in size to a modern day chimp), being only a fifth the size of human's. Hence, **bipedal locomotion** predates evolution of a large brain in our ancestry. While *Ardipithecus* is believed to have lived in wooded areas, the next stage of evolution, *Australopithecus* ("southern ape" – various species), which dates back to 4.2 million years ago, was more of an open savannah dwelling creature. *Australopithecus* existed in two main forms – a robust (strong and full-bodied) and a gracile (more delicate) form, and both had a cranial capacity or around 450 cc. Although this is a little larger than *Ardipithecus,* it is still only 35% the size of modern humans. Around 2.5 million years ago the first of the

KEY TERMS

Hominid and **hominin** *Hominid* refers to all existing and extinct great apes and modern humans. In contrast, *hominin* refers to modern humans and all of our immediate ancestors, such as *Australopithecus* species and *Homo habilis.*

Bipedal locomotion The ability to walk upright on two legs rather than with quadrupedal locomotion (i.e., walking on four legs).

Homo line – *Homo habilis* ("handy man") evolved from the gracile form of *Australopithecus. Homo habilis* was so named because it is believed that it was the first of our ancestors to make use of (quite primitive) stone tools. Such tools included scrapers to remove meat from the bone (*H habilis* is known to have included meat in its diet). Experts today suggest that *H habilis* used primitive stone tools to scavenge carrion rather than to hunt, but this is, of course, difficult to determine. The brain capacity of *H habilis* was around 600 cc, which is somewhat larger than *Australopithecus* – suggesting perhaps that tool use requires a larger brain (see below). Although its body was still ape-like, it had less of a protruding face than previous ancestors.

Sometime around 1.8 million years ago, *H habilis* gave rise to another species, known as *Homo erectus* ("upright man"). *H erectus*, having a much larger brain capacity (850 cc) than that of *H habilis*, certainly did engage in hunting and in order to do so developed more sophisticated tools. There is also evidence that members this species were able to make use of fire. While, like *H habilis* before, *H erectus* originated in Africa, and it was the first of our ancestors to leave that continent. In fact, during the period between 1.8 million and 300,000 years before present, they spread throughout Asia as far east as Java and through much of Europe as far west as Britain. Some experts consider that *H erectus* was the first of our ancestral species to live in small hunter–gatherer societies, where the old and infirm are provided for (Boehm, 1999).

At some stage around 400,000 years ago *H erectus* gave rise to early *Homo sapiens* ("thinking man" – sometimes called "archaic humans"). Such early humans had a brain capacity of around 1,000 cc, hence a considerable increase when compared with *H erectus* but still at least 20% smaller than current humans. If you bumped into an archaic human today you would clearly recognise them as human but would think them a bit intellectually slow no doubt. A number of subspecies arose from this early *H. sapiens,* such as the Neanderthals, referred to above. These are today classified as *Homo sapiens neanderthalensis* (although just to confuse matters some, authorities classify them as their own species of *Homo neanderthalensis*). The Neanderthals were short and stocky, with heavy eyebrow ridges and huge noses. They are often perceived as a brutish sideline that appeared briefly in the fossil record before becoming extinct. Given, however, that they lived from around 300,000 up until perhaps 20,000 years before present, and given they spread throughout much of Europe and the Middle East, they can certainly be seen as a successful subspecies. Indeed, their robustness was probably an adaptation to Ice-Age conditions rather than a throwback to our more ancient anthropoid ancestors. Their brains were of equal capacity to modern day *Homo sapiens,* at around 1,450–1,500 cc on average.

DISCUSS AND DEBATE – BIPEDAL LOCOMOTION

Make a list of the advantages and disadvantages of bipedal locomotion and of quadrupedal locomotion.

Why might larger brains have evolved following the evolution of bipedalism?

Why haven't other great apes developed full bipedal locomotion?

Truly modern humans (*Homo sapiens sapiens*) began to emerge somewhere between 100,000 and 200,000 years ago, and current living humans have changed very little anatomically since this period. An example of early modern humans is the Cro-Magnon people, referred to above.

KEY CONCEPTS BOX 2.2

Multiregion versus Out-of-Africa hypotheses

Currently there are two main hypotheses to explain the emergence of *Homo sapiens sapiens* (*Hss*) – the **multiregion** and the **Out-of-Africa Replacement hypotheses**. According to the multiregion hypothesis, *Hss* evolved from *H erectus* in a number of different geographical locations around Eurasia, whereas for the Out-of-Africa Replacement hypothesis all modern human populations are derived from a small number of individuals that lived in Africa around 150,000 years ago and rapidly replaced earlier populations, such as the Neanderthals. Recent genetic evidence gathered from small, energy-producing bodies in our cells, called **mitochondria** (each of which has its own small number of genes passed down only through the female line), would favor the Out-of-Africa Replacement hypothesis. It has even been suggested that all present living humans can trace their ancestry back to one female living in Africa around 150,000 years ago – **Mitochondrial Eve**.

KEY TERM

Mitochondrial Eve The most recent female ancestor of all living humans. Lived in East Africa approximately 190,000–200,000 years ago. Also know as *African Eve*.

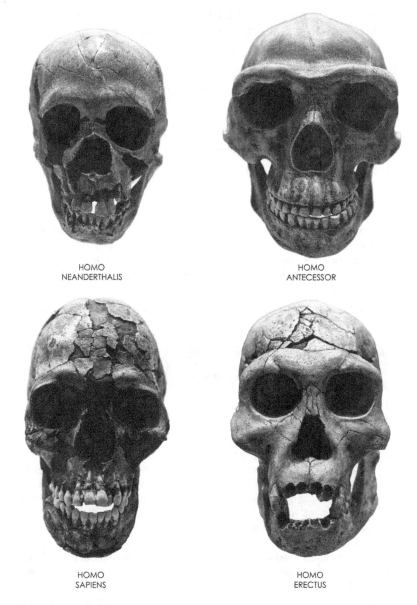

HOMO
NEANDERTHALIS

HOMO
ANTECESSOR

HOMO
SAPIENS

HOMO
ERECTUS

Figure 2.1 Hominin cranial capacity.

Source: Image copyright © Creativemarc/Shutterstock.com.

Once we were not alone

Throughout this description of human evolution we have emphasized the increase in brain size. We need to be careful, however, not to assume that this was in some way predestined – for each of the prehuman species the brain size was appropriate for its purpose. It is only when the environment changes that adaptive change then occurs (see below). Large brains also have costs. They use up a lot of energy (in humans, around 25%), meaning we need more food; also, big-headed babies make birth dangerous to mother and child. We should also avoid falling into the trap of perceiving human evolution was a straight line passing through a geological time-frame, with *Ardipithecus* at one end and *Homo sapiens sapiens* at the other. The story of our evolution is one of branching change, with many species appearing and disappearing. This means that while a new species may have taken a slightly different path as it evolved adaptations to suit the local environment, members of the original species will often have continued to have existed in other areas. Hence, many of the proto-human species described above will have coexisted and may even have come into contact. (In fact there is now clear evidence that humans and Neanderthals did come into contact).

The evolution of *Homo sapiens* is characterized by the development a number of features that distinguish us from the great apes, including upright walking posture (bipedal locomotion), a great reduction in the growth of body hair and most importantly a greatly enlarged brain. All of these changes are considered to be adaptive ones – that is, they came about through Darwinian selective forces to provide us with survival and reproductive advantages. In particular, the increase in brain size and accompanying increase in intellectual ability has intrigued evolutionists from Darwin right up to present day evolutionary psychologists. So, given the costs of running this large organ (see above), what were the advantages of evolving a brain the size of a grapefruit, given we got on fine with one the size of a lemon a mere 4 million years previously?

Evolution of brain size and intelligence

Clearly, the evolution of a larger brain went hand in hand with the intellectual abilities that it supports, such as language, tool use and complex social skills. Given, however, that the primates we evolved from were already intelligent when compared to other mammalian species, and given that we alone demonstrated rapid brain expansion, this raises the question, what was it about our evolution that led to a level of intellect that now allows us to dominate the planet? Currently two main types of competing theories have been developed to explain the increase in human brain size/intelligence relative to other primates – those based on ecological factors and those based on social factors. Both draw on existing primate adaptations.

Ecological explanations of human intelligence

Ecological theories of human intelligence place the problems of efficient foraging at center stage. Due to this, such hypotheses are often labeled *ecological demands explanations*. Early primates were largely fruit-eating (**frugivorous**) inhabitants of trees (**arboreal** living). This means that the demands of extracting fruit from the three-dimensional environment of the forest canopy is seen as the major initial ecological pressure that led to the evolution of a large, intelligent brain. Hence, natural selection favored those individuals that were most able to overcome the navigational challenges involved in arboreal foraging. This, of course, only helps to explain why primates are intelligent and not why our human ancestors continued to develop this further. Had there not been a change in equatorial Africa around 14 million years ago, we probably would have not have. At this point, in our evolutionary past, due to climatic changes in Africa, however, the open grasslands that we now call the savannah began to enlarge as the forests shrank. This opened up a new ecological niche for our ancestors as they split into two groups – apes that remained semiarboreal (ancestral chimps) and those that increasingly became adapted to savannah dwelling (hominins such as *Ardipithecus* and subsequently *Australopithecus,* followed, in turn, by the *Homo* line). Hence, in a sense, it might be argued that human intelligence ultimately comes down to grass.

In coming out onto the open savannah, so the argument goes, our hominin ancestors found new uses for their arboreal adaptations. Examples of these include good gripping hands (originally put to use grabbing branches and handling fruit) and effective stereoscopic color vision (originally used to determine the ripeness of fruit). Both of these features were now put to good use to aid scavenging and hunting. Also, the increase in bipedal locomotion allowed our early ancestors to see over the long grass. As we have seen between 2.5 million and 200,000 years before present, our ancestors developed increasingly sophisticated tools for hunting and scavenging. Note that the ability to copy others in the preparation and use of such tools may also have been part-and-parcel of the ecological demands explanation for the evolution of human intelligence (see Chapter 9). Interestingly, while some authorities, such as Robert Ardrey (1970) emphasized the demands of hunting as the main evolutionary impetus for the development of a larger brain, others have highlighted the intellectual demands of gathering plant food as the primary driving force (Tanner, 1981). Both of these, however, can be considered as ecological explanations of human intelligence.

Social complexity explanations of human intelligence

Ecological theories for the evolution of human intelligence held sway for much of the 20th century. Beginning in the mid 1970s, however, a new driving force began to be considered – social complexity. In 1976, University of Cambridge theoretical psychologist Nick Humphrey was the first to observe that the most intelligent species are also the most social ones (think, for example, of elephants, dolphins and chimpanzees). To Humphrey, the intellectual problems of the physical environment pale into insignificance when compared with those that require complex social interactions (Humphrey, 1976). Others took on board Humphrey's arguments, and in 1988 St. Andrews animal behaviorists Andy Whiten and Dick Byrne brought these ideas together in a book called *Machiavellian Intelligence* (Byrne & Whiten, 1988). This proposed the **Machiavellian intelligence hypothesis** for the development of intelligence throughout the animal kingdom – but also emphasised the advantages of having such abilities for early hominins (Whiten & Byrne, 1988).

> **KEY TERM**
>
> **Machiavellian intelligence hypothesis** The notion that advanced cognitive processes arose in primates and hominins as adaptations to deal with social complexity rather than non-social challenges, such as finding food or developing tools.

KEY CONCEPTS BOX 2.3

The origin of the term *Machiavellian*

The term *Machiavellian intelligence* is derived from an Italian political treatise called *The Prince*, written in the 16th century by Niccolò Machiavelli (1988). In *The Prince*, Machiavelli presented the basic argument that to gain and hold onto power a leader should not be swayed by moral codes imposed from society but rather must use whatever means are most effective (pleasant or unpleasant – see King, 2004).

Moreover, Machiavelli also suggested that a leader should look into the minds of both supporters and competitors in order to manipulate them. Although Machiavelli believed that when leaders are getting their own way they should treat others with benevolence, ever since the publication of *The Prince* the term *Machiavellian* has been used as a term to describe devious political maneuvering.

Whiten and Byrne's Machiavellian intelligence version of the social complexity hypothesis suggests that the ability to predict the behavior of others provided a selective advantage in that it allows for individuals to engage in "adaptive social maneuvering." In relation to evolution, this means that individuals that happened to have the genes to develop such intellectual complexity would then be able to manipulate others to their own advantage. Even a slight ability to do this would have great advantages, especially when it comes to deception over those that lack this ability. But such an ability to deceive requires a large increase in brain power – which is why it is suggested as the driving force for evolution of intelligence. Signaling the truth to another individual is quite easy – you simply report, for example, your internal state – such as whether you are feeling hungry or amorous. Deception, however, requires some ability to consider the state of mind of the other individual (to pretend to be hungry, for example, you might have to consider whether the other individual was aware you had eaten recently). This ability probably arose in our primate ancestry in some very simple form – but in terms of evolutionary success, it proved to be dynamite. A good way to imagine the advantages the first individuals capable of deception would have had over others is to consider the 2009 feature film *The Invention of Lying,* starring Ricky Gervais and Jennifer Garner. Set in an alternative reality, the character played by Gervais, Mark Bellison, through some accident of nature, becomes the only person on the planet able to tell a lie. Within weeks he becomes rich, famous and suddenly more attractive to the glamorous Anna (played by Garner), whom he desires. While *The Invention of Lying* is no doubt quite a long way from early primate Machiavellian behavior, it does have parallels with aspects of recorded chimpanzee behavior. In 2009, at a zoo in Stockholm, for example, a chimp called Santino was regularly observed gathering and hiding rocks in his enclosure that he later threw at visitors that came to watch him. The fact that he gathered them prior to the arrival of visitors and that he hid them suggest he was using **planned deception** – a form of behavior that we normally only associate with people. Many field observations of chimps suggest that, in addition to planned deception, they are quite capable of social maneuvering both companions and competitors (Pearce, 2008).

How the original ability to glean some degree of insight into another's state of mind came about is a matter for conjecture and debate for those that support this version of the social complexity hypothesis. More than likely it was a random genetic mutation (see later) that led to some new neural circuitry. But what is not in debate is the fact that, given natural selection is based on competition between individuals, then over time such an ability would quickly spread throughout a population as it favored those most able to predict the behavior of others. Then, according to this form of the social complexity hypothesis, over many generations on the open savannah as group size increased, so did the neural hardware required to support this ability (again supported by selection of a number of random genetic mutations leading to small incremental steps towards greater complexity).

KEY CONCEPTS BOX 2.4

Dunbar's number and the social grooming hypothesis

Have you ever considered just how many people you know? According to University of Oxford evolutionary psychologist Robin Dunbar, there is an upper limit that most of us reach in life – 150. Dunbar noted that primates maintain social bonds by grooming each other and that those belonging to larger groups have both larger brains and spend increasingly longer grooming each other. To Dunbar (1993), as our social group increased we shifted from grooming physically to grooming with language. Since we can groom several people by speaking, this shift in how we retain an attachment to friends freed up more time to engage in other activities. But how did Dunbar arrive at 150 as the upper limit? There are two forms of evidence that support this. First, Dunbar worked out that the size of our **neocortex** (the most recently evolved outer covering of our forebrain – sometimes called the *social cortex*) suggests we can only know that many people in any sort of detail. And second, time and again, the figure of around 150 crops up in group sizes found on the planet from the size of hunter–gatherer bands, to religious groups, to the number of people we have on our Christmas card lists and even the number of (real) friends on Facebook. This figure has subsequently become known as **Dunbar's number,** and what we can call his **social grooming hypothesis** is a version of the social complexity hypothesis for the evolution of the human intellect.

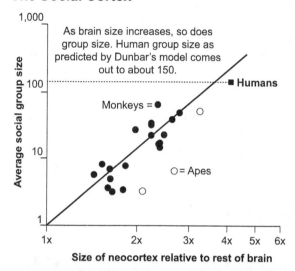

Figure 2.2 Dunbar's number and the social grooming hypothesis.
Source: Adapted from Dunbar (1993).

How well do the ecological and social complexity hypotheses stand up to scrutiny?

As with evolutionary theories that are formulated to explain behavior in general, because they deal with the ancient past, we can only gather evidence indirectly about these hypotheses. Can we, however, for example, identify scientific findings that would be consistent with these hypotheses (or, importantly, can refute them)? The ecological hypotheses are supported by the fact that animals that have to process their food (such as determining the ripeness of fruit and then peeling it) generally have larger brains than those that simply graze. Moreover, it is certainly the case that there is a close relationship in the fossil record between increases in the complexity of tools developed and increases in brain size during our evolution. For the social complexity hypotheses there is also strong correlational supportive evidence because there is a positive relationship between the size of the group a given species of primate lives in and the size of its neocortex (see Key concepts box 2.3). Two points can be made about these findings. First, of course, correlational evidence does not prove a causal relationship. Having a large brain and forming large groups (or developing complex tools) may all be the result of some other evolutionary driving force. Hence, we have to be careful with correlational evidence. Second, the ecological and social explanations for the development of a large intelligent brain might not be mutually exclusive – they might both have contributed to a degree. This is a question we will return to in Chapter 7 when we consider the relationship between evolution and cognition.

DISCUSS AND DEBATE – EXPLANATIONS FOR EVOLUTION OF INTELLIGENCE

Compare and contrast ecological and social complexity hypotheses for the evolution of human intelligence.

Consider how many people you know – does this fit in with Dunbar's number? You might want to consider how many friends you have on Facebook (and how many of these "friends" are real friends!)?

Mendelian genetics – The physical foundation of evolutionary change

At the time of Darwin's death in 1882, belief in his notion of evolution by natural selection had dropped significantly in the "opinion polls." The problem for Darwin was that while he had developed a serious contender for an explanation of evolutionary change, his theory lacked a physical mechanism of inheritance. Darwin was unaware of the existence of genes. The problem for natural selection was the belief that any new favorable features that arose would quickly be diluted over a number of generations as it was widely believed that inheritance involved blending, and "watering down," of characteristics. Hence, how could that cornerstone of natural selection – selective advantage – be maintained? As an example imagine that being tall would confer an advantage in a given environment (e.g., picking fruit from higher branches). According to the prevailing view, when this feature arose over several generations, this advantageous characteristic would simply dissipate due to the blending of characteristics pushing height back to the population mean. Ironically, this problem was solved during Darwin's lifetime with the discovery of genes – it's just that nobody realized this. The research had been done by Austrian monk Gregor Mendel, who demonstrated that characteristics don't blend. Mendel had even published his findings – but in a small local journal (results in the *Journal of the Brno Natural History Society* in 1866), and the importance of the findings had not been realized.

Mendel's laws

Mendel studied inheritance in garden pea plants at Augustinian Abbey gardens of St. Thomas in Brno (now in the Czech Republic) between 1856 and 1863. During this period he conducted a series of painstaking breeding experiments involving almost 30,000 plants. By carefully using pollen to self-fertilize and cross-pollinate, he was able to study the inheritance of distinct alternate characteristics, such as plant height or color, of peas or flowers. Mendel made three important discoveries (**Mendel's laws of inheritance**) that eventually led to the development of the new science of genetics. First, he demonstrated that **traits** (characteristics such as pea color) are coded by genes acting in pairs (one from each parental sex cell, or **gamete**). Second, he showed that the relationship between physical traits (the **phenotype**) and the underlying genes (the **genotype**) is not straightforward since two plants can have the same phenotype but a different genotype. The reason for this is because one of the genes in each pair can be dominant and the other recessive. In the case of dominant genes, only one copy is necessary for the trait to be expressed, whereas for recessive genes two copies have to be present (see hair color in Key concepts box 2.5). By tradition, the dominant **gene** is given a capital letter and the recessive one a lower case letter (e.g., B and b for human eye color). When an individual has two copies of a dominant or a recessive gene, it is said to be **homozygous** (*homo* means "alike" – hence, BB and bb – see below) for the trait in question, whereas if it has a mixture of dominant and recessive genes it is said to be **heterozygous** (*hetero* means "different" – hence, Bb). Third and finally, Mendel showed that genes do not blend together but are "particulate" and are passed on intact. Hence, in the case of his breeding experiments, peas were either green or yellow, with no blended intermediates.

These findings really did solve Darwin's problem of how selected characteristics would not be watered down over generations but continue to be passed down due to the advantages they confer. Unfortunately for both Mendel and Darwin, these results did not make it into mainstream science until the 20th century, when they were rediscovered and integrated into Darwin's theory to form what became known as the **modern evolutionary synthesis** (of genetics and natural selection), or **Neo-Darwinism**.

KEY TERM

Neo-Darwinism Darwinian evolution that is informed by Mendelian genetics. Also called the **modern evolutionary synthesis** or simply the **modern synthesis**.

KEY CONCEPTS BOX 2.5

The Punnett square and human eye color

In 1910, the Cambridge geneticist Reginald Punnett devised a simple way of illustrating the potential genotypes that we can result from various matings – the **Punnett square**. If we look at human eye color for example, brown eyes (B) are dominant to blue eyes (b). Imagine if a heterozygous brown eyed man (Bb) marries a blue eyed woman (bb). Knowing that the mother's genotype is bb and the father is Bb allows us to create a simple Punnett square that demonstrates the genotype ratios in any offspring they might, have thus (see Figure 2.3):

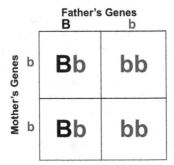

Figure 2.3 The Punnett square and human eye color.

Note that in this case, 2/4 (50%) of the offspring are likely to have the genotype Bb, which codes for the phenotype of brown eyes, and two are likely to have the genotype bb and will have blue eyes. If the father had been BB, then all of the offspring would have had brown eyes (all would be Bb, and you only need one copy of a dominant gene for the trait to be expressed).

Developments in genetics since Mendel

Although Mendel's "laws" of genetics are considered broadly correct today, in the 150 years since he published his original work, genetics has developed into a highly complex science. Since the 1930s, it has been known that genes are found at specific locations (**loci**) on long rod-like paired bodies called **chromosomes,** which are found in the **nucleus** of each cell. Humans have 23 pairs of such chromosomes (one of the pair from each parent), and a person's entire genetic

make-up is known as her **genome**. In 2003,
the **Human Genome Project** sequenced the
entire human genome – hence we now know
the location of every human gene at each
locus for every chromosome. Each specific
locus can generally have more than one
alternative form of a gene. These alternate
forms of a gene are called **alleles**. Hence
B and b are alleles of the eye color genes coding for brown and blue,
respectively. (Incidentally, today it is known that a number of other
genes can alter precise eye color, including G for green or hazel and g
for light colored eyes. Given that these are found at a different locus,
we can have a large range of eye coloration with a large number of
genotypes, such as BbGg and BBgg. To have true blue eyes, you have
to have the gene combination of bbgg – two pairs of recessive genes.)

Mendel was lucky in choosing to make use of pea plants because
the inheritance of their characteristics is quite straightforward. In fact
Mendel was lucky in at least three different ways. First, in the characteristics he studied, the genotype–phenotype relationship was really
quite straightforward. That is, the traits he focused on were coded for
by simple gene pairs, a situation that is labeled **monomorphism**.
Many traits are determined by more than one type of gene which is
called **polymorphism** (eye color discussed above is an example of a
polymorphic trait). Second, Mendel was also fortunate in that dominance of a trait is not always "complete." A dominant gene that always
leads to that trait occurring in the population is called **complete penetrance**. Complete penetrance occurred in the features Mendel was
interested in in his pea plants. In many cases, however, the dominant
gene does not express itself in all of the members of the population that
have it – a situation called **incomplete penetrance** (the language of
genetics is all very logical). Incomplete penetrance may be due to one
of two reasons – genetic and environmental. In the case of genetic
causes of incomplete penetrance, other genes have an effect that
modifies that particular gene, whereas in the case of environmental
causes things like diet and trauma may water down the effects of the
dominant gene (interestingly, for some of us eye color fades as we
age). Finally, Mendel was fortunate in that, in addition to many traits
requiring more than one gene, some genes can affect more than one
trait – a situation that is known as **pleiotropy**. An example of pleiotropy
in humans is **albinism,** where a single mutant gene has a number of
phenotypic effects, including very pale skin and poor vision.

Polymorphism, incomplete penetrance and pleiotropy are three
features of genetics that demonstrate how complicated a science it

has grown up to be. These caveats are, however, only the beginning of just how complex the relationship between genotype and phenotype really is, especially when we are dealing with human psychological abilities (see later). Before we can begin to examine that relationship, we need to know a little more about genetics – such as what exactly is a gene and how can it possibly affect behavior and our state of mind?

Molecular biology – What is a gene, physically?

Molecular biology deals with the structure of biological entities, such as genes and proteins. It really began to develop with Watson and Crick's 1953 discovery of the structure of **DNA.** To a molecular biologist, a gene is a portion of DNA that codes for the production of a large molecule called a **polypeptide**. A polypeptide is either a protein or, in the

> ### KEY TERMS
>
> **DNA (Deoxyribonucleic acid)** The macromolecule that contains all of our genetic information encoded in a series of base pairs.
>
> **Polypeptide** A small protein or part of a larger protein. Coded for by a gene.

case of many larger proteins, part of a protein (some proteins are made up of 1, 2, several or many polypeptides). Genes are really a sequence of units that make up the famous twisted ladder that is the double helix discovered by Watson and Crick.

Each unit consists of three parts – the "rails" or back bone of the ladder being alternating acid and sugar portions and the rungs consisting of one of four different base pairs: **adenine** and **thymine**, **cytosine** and **guanine** (A, T, C and G – see Figure 2.4). *A* can only pair with *T*, and *C* can only pair with *G*, and each rung is in effect made up of a base that juts out from each side and makes a weak bond with its partner base. When a protein is required to be manufactured, these weak (hydrogen) bonds break and portions of the helix unzip to reveal two sequences of bases, such as AACTTCAGG. The sequence on one side is used as a recipe to form a polypeptide (via a sequence of reactions involving other molecules). How this occurs involves other molecules before the protein is complete – but fortunately for our purposes we do not have to explore the molecular basis of how genes operate further. (Those who wish to know this in great detail should see the Further Reading section.) The entire sequence of these base letters on all of our chromosomes is what the Human Genome Project revealed – in all 3 billion base pairs. (If you unraveled the entire DNA from a single human cell, despite being a molecule, it would stretch out for 1.5 meters in length – truly a **macromolecule**.)

Note that the gene is both a portion of the giant double helix molecule DNA and an instruction code to make part of or all of another molecule (it also makes copies of itself). It used to be thought that

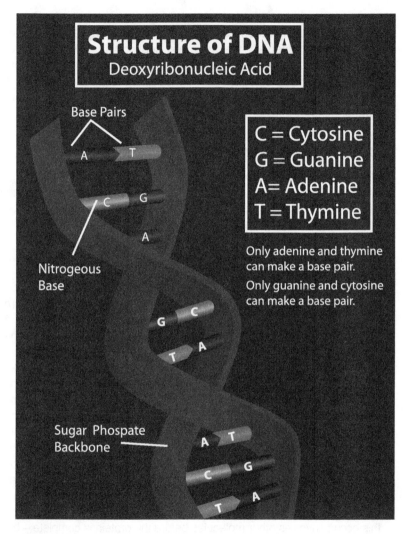

Figure 2.4 Structure of DNA.
Source: Image copyright © iris wright/Shutterstock.com.

genes, in coding for the production of proteins on a chromosome, are simply like beads on a necklace, arranged in a lengthy row along the chromosome. Today it is known that this is, however, a bit of a simplification. Since protein production generally requires the gathering together of information that is distributed over many parts of the chromosome it requires the decoding of different bits of DNA from many parts of this macromolecule. So perhaps a better analogy would be a computer hard drive that takes information spread out from all over the disk to conduct a task.

What do proteins do?

Because the function of proteins depends, in part, on their shape, their name is derived from the ancient shape-changing Greek sea-God **Proteus**. Proteins do many things, from acting as enzymes and chemical regulators in the body to transporting substances across cell membranes. As far as we are concerned, however, the most important job that proteins do is to form much of the structure of the brain and the **neurotransmitters** that send messages between **neurons** contained within. Nearly all body cells contain all of our genes, but which genes are expressed (i.e., active in that area) varies from one part of the body to another. The human brain is a 1.5 kilogram mass of proteins and fats, and, astonishingly, about a third of our genes are expressed in production and maintenance of this organ (far more than in any other part of the body). Since differences in intelligence and personality between people are believed to be determined, in part, by differences in their level of neurotransmitters (and by differences in their **receptor sites** on neurons for such neurotransmitters), and since proteins are coded for by genes, then this is why people can differ due to differences in their genes. In the next section we will briefly examine this relationship between genes and behavior.

From genes to brain and behavior – Three timescales

It is regularly reported in the media that scientists have discovered a gene that leads to quite specific human behavioral traits – such as a "gay" gene or a gene for intelligence. This, of course, is a gross simplification of the relationship between genes and behavioral responses. It is important to realize from the outset that genes do not code for behavior. It is, however, accurate to say that *differences* between people in their genome can contribute to *differences* in their behavior. (Note that to say that a gene may have been identified by scientists that might contribute to differences between people, in their sexuality or intelligence, sounds a bit convoluted and does not make for a great newspaper headline.) Although a detailed understanding of the complex relationships that connect genes, brains and behavior would be out of place in this book, it is necessary to have a broad grasp of this relationship.

Genes regulate the development and organization of the brain within a developmental and an evolutionary context. This means that the relationship between genes, brain and behavior takes place over three quite different timescales (Robinson, Fernald & Clayton, 2008). First, (genetically influenced) brain activity leads to behavior on an immediate moment-by-moment timescale. An example of this might be when you have a face-to-face conversation with a friend (people used to do this before cell phones were invented). Such activity involves a number of areas of the brain (think about it – you have to be able to identify the friend, get close enough to gather their attention and then use several areas of the brain in order to compose and produce sentences and to listen to and understand their sentences). Second, over an individual developmental time frame, the genetic code, through interaction with environmental input, influences brain development which, in turn, influences behavior. An example of this might be the neural circuitry that was laid down during development to support language so that you can talk to your friend. Note that while we can consider the ability to speak as an evolved adaptation, the actual words and grammar used will be culturally determined (see Chapter 7). Finally, over an extremely lengthy evolutionary timescale, the process of natural selection modifies the genetic code in a species to influence the type of brain that individuals of that species will develop. Sticking once again with language, it is clear that evolutionary processes led to the type of brain that members of our species now have that facilitates the development and use language.

DISCUSS AND DEBATE – THE HUMAN GENOME PROJECT (HGP)

Prior to the sequencing of the human genome, it was assumed that humans had at least 100,000 genes. The HGP has, however, discovered that we only have a little over 20,000 genes. Does this finding alter the way that we perceive ourselves?

Heritability and behavior

Heritability, which we first encountered in Chapter 1, is an important concept in genetics. Technically, heritability is defined as the extent to which variation between individuals in a population is due to genetic (rather than to environmental) differences between them. Note that heritability is not a measure of what proportion of a trait is caused by genes

and what proportion is caused by the environment. Once again, the key word here is *differences*. Hence, we are dealing with an estimate of the proportion of differences in a trait between individuals that is due to their genetic differences. Most geneticists that make use of the concept of heritability conduct research on physical traits, such as meat and crop production. Such applications are rarely seen as controversial. In fact, knowing the extent to which productivity can be improved through selective breeding has been of great benefit to agriculturalists. Even when considering the heritability of behavior in animals the concept of heritability is rarely controversial. Dog breeders, for example, have been able to selectively breed for temperamental features, such as placidity in house pets – which turns out to be highly heritable (Jones & Gosling, 2005). When, however, the concept is applied to human personality and intelligence, then impassioned argument invariably arises. In particular, the notion put forward by some evolutionists, that intelligence is highly heritable, has led to a passionate nature–nurture debate. The field that deals with the relationship between genes and behavior is known as **behavioral genetics** – a controversial area that we consider further in Chapter 6.

> **KEY TERMS**
>
> **Heritability** An estimation of the extent to which the phenotype is influenced by the genotype for a given trait.
>
> **Behavioral genetics** The field of study that examines the role of genes in behavior in humans and animals. Draws on a number of other fields of science, such as genetics, psychology and statistics.

In this chapter we have outlined a number of areas where our knowledge of evolution has improved dramatically in recent years. Such areas include human evolution, including physical adaptation and tentative notions of evolutionary driving forces for intellectual development. We have also introduced the field of genetics and the controversial concept of heritability.

Summary

- The human lineage split from the apes sometime around 5–6 million years ago. Several "stages" of human evolution have been identified, including *Ardipithecus* and *Australopithecus*, and *Homo habilis*, *Homo erectus*, *Homo sapiens* (archaic) and, finally, anatomically modern *Homo sapiens sapiens*. During this period, our ancestors began to walk upright (bipedal locomotion), and this was followed by a rapid enlargement of the human brain (expanding from 300 cc to 1,500 cc).

- Two main types of hypothesis have been developed to explain the evolution of this rapid expansion in size and growth in intellectual ability that accompanied it – **ecological** and **social complexity** hypotheses. *Ecological hypotheses* suggest that the main driving force for our expansion of brain and intellect was the improved ability this gave our ancestors in gathering food and later on in making tools for hunting. In contrast, *social complexity hypotheses* suggest that we evolved more complex brains in order to negotiate the challenges of living in an increasingly large social group. An extreme example of this – the Machiavellian intelligence hypothesis – suggests that being able to predict and manipulate others was the **selective advantage** to our early ancestors. According to Dunbar's number we can only really know up to 150 people since this is the upper limit of the amount that the neocortex of our brain has evolved to deal with.
- Mendel provided the missing piece of Darwin's theory of evolution by natural selection by uncovering three **"laws of inheritance."** This led to the development of the field of **genetics** during the 20th century. Genes consist of portions of the DNA (deoxyribonucleic acid) that make up our chromosomes. A gene codes for proteins, which then form a multitude of important biological substances, such as enzymes and much of the human brain and its neurotransmitters. Due to the Human Genome Project, it is known today that humans have approximately 20,500 genes, with around one third of these being active (**expressed**) in the brain.
- The relationship between genes and behavior is highly complex and always involves interaction with environmental input. Genes do not code for behavior, but differences in genes (alleles) can contribute to differences in behavior between them. This relationship takes place over three timescales – first, genetically influenced brain activity leads to behavioral outcomes; second, during an individual's lifetime the brain develops within an environmental context, leading to appropriate behavior; third, the form of brain that our species has today was influenced by natural selection during our evolutionary past.
- *Heritability* consists of the extent to which variation between individuals in a population is due to genetic differences between them. The concept of heritability is a contentious one when applied to human behavioral traits, such as intelligence and personality.

FURTHER READING

Barkow, J. H., Cosmides, L. & Tooby, J. (Eds.) (1992) *The adapted mind: Evolutionary psychology and the generation of culture.* Oxford: Oxford University Press.

Dunbar, R. (2014). *Human evolution: A Pelican introduction.* Pelican: London.

Nettle, D. (2009). *Evolution and genetics for psychology.* Oxford: Oxford University Press.

Pallen, M. (2009). *The rough guide to evolution.* Penguin: London.

Darwin's second selective force

Sexual selection

3

What this chapter will teach you

- How Darwin's second evolutionary force – sexual selection – has been refined.
- The advantages and disadvantages of sexual and asexual reproduction.
- The Red Queen hypothesis to explain why sexual reproduction can be superior to asexual reproduction.
- How sexual selection has been used to help explain human intelligence and sex differences in aggression.

In this chapter we return to Darwin's second selective force – **sexual selection** – and consider why it is that many organisms use the bizarre form of reproduction that we call *sex*. We also present and explain the Darwinian lexicon of sexual selection theory. We then consider an influential hypothesis that relates sexual selection to the evolution of human language and intelligence. Finally, we consider how sexual selection theory has been used to help explain sex differences in aggressive behavior.

Sexual selection – From Darwin to the present

You may recall from Chapter 1 that when Darwin considered the gaudy features of many male birds and mammals, such as the tail feathers of the peacock or the red face and "mane" of the male mandrill, he was not convinced that these augmentations could have arisen from **natural selection.** This is also true of the fact that the males of many species engage in more physical displays than the females (Darwin, 1871). The problem was that because natural selection was believed to lead to features that aid survival (see Chapter 1), how could characteristics that make individuals stand out so vividly to predators be sustained in a population? Surely, more drab males would have an advantage over the colorful, exhibitionist males in avoiding predators? In fact, while the notion that these features form an encumbrance to survival was mere (educated) speculation on Darwin's part, today it is known that he was certainly right. The male peacock, for example, is taken by predators in greater numbers than his drab female counterpart. Moreover, it is precisely these extravagant tail feathers that increase the likelihood of tigers (historically the main predator of peafowl) capturing them. In order to explain how these conspicuous features evolved, Darwin introduced a new concept – sexual selection.

As we saw in Chapter 1, sexual selection involves both male–male competition and female choice. Darwin was aware of this and considered that we can break sexual selection down to two components – **intrasexual selection** and **intersexual selection**. *Intrasexual* means "within sex" and is concerned with competition between members of one sex for access to the opposite sex. In contrast, *intersexual selection* means "between sex" and is concerned with impressing the opposite sex. Darwin described intrasexual selection as the power to conquer other males in battle and intersexual selection as the power to charm the females. Although these descriptions may sound a little sexist today, many studies have confirmed that this is the way it generally works in the animal kingdom – with males competing between themselves and doing so for female attention (Gould & Gould, 1997).

This raises the question – Do human males and females also differ along these lines? This is a question we will consider in Chapter 4, after first exploring Darwinian terminology and examining developments since Darwin's day in this chapter.

KEY TERMS

Intrasexual selection Competition between members of one sex for access to members of the opposite sex. Generally leads to a lower threshold for aggression in males of a species.

Intersexual selection Competition to attract members of the opposite sex. Generally leads to "choosy" females and "sexy" (sexually selected ornamented) males.

Figure 3.1 Male mandrill.

Source: Image copyright © Nagel Photography/Shutterstock.com.

Darwinian terminology – Primary and secondary sexual characteristics

In developing sexual selection theory, Darwin introduced a whole series of terms into biology that are still in use today. He described sex differences that are directly related to reproduction as **primary sexual characteristics** (that is, the reproductive organs). In contrast, features which have evolved to enable competition or courtship (but have no direct role in reproduction) he called **secondary sexual characteristics**. Examples of this might be the peacock's tail feathers or the antlers of many species of deer. Note that the tail is an example of what Darwin called an **ornament** (it serves the purpose of attraction) whereas the antlers he described as **weapons**. Note also that as is the case with the deer antlers, it is possible for a feature to be both a weapon and an ornament. Finally, Darwin introduced the term **sexual dimorphism** to describe differences between the sexes that arose from sexual selection (i.e., secondary sexual characteristics). These differences can be both structural (e.g., the size difference between males and females of many species) and behavioral (e.g., the lower threshold for aggression observed in the males of many species). Whereas most social scientists today explain behavioral differences between men and women as the product of learning to conform to gender roles ascribed by a given society, evolutionary psychologists generally find this explanation wanting and emphasize instead the role played by sexual selection (Buss, 2003, 2013b). We will return to this point later on when we consider aggressive behavior.

> **KEY TERMS**
>
> **Primary sexual characteristics** Differences between males and females that have evolved for reproductive purposes, such as the uterus and the penis.
>
> **Secondary sexual characteristics** Differences between males and females that are not directly related to reproduction but rather evolved to help gain access to the opposite sex. These usually involve exaggerated features in males, such as ornamental feathers, or weapons, such as large teeth or horns.
>
> **Sexual dimorphism** A general term introduced by Darwin to describe physical and behavioral differences between the sexes.

Fisher's runaway selection – Living fast and dying young

In 1915, English geneticist Ronald Fisher proposed an explanation for just why sexually selected displays often became quite so elaborate in males. Fisher suggested that whichever feature was to become elaborate began as a signal of genuine fitness to females (i.e., fitness in the sense of being useful for survival and reproduction). Staying with the example of the peacock's tail feathers, in the distant evolutionary past a male with good tail feathers that helped to support flight would have made a good choice for the females of his species. Over time, due to this female preference, male tail feathers would gradually become larger and more colorful in order for these qualities to stand out from the crowd. But, simultaneously, females with this preference would also become more numerous (because females choosing fitter males would have more offspring who would also tend to inherit this trait). Eventually, as the tail feather become more and more elaborate (and the females consistently chose the larger, brighter ones), the characteristic would "run away" from its original function (improved flight due to natural selection). Hence, according to Fisher, female choice can lead to characteristics in a male running away from their original function and becoming an encumbrance – hence, **runaway selection**. To Fisher this didn't matter, as long as the male was successful in attracting females. It didn't even matter if his life was shortened, provided he produced more surviving offspring during that short life than those more drab males did (during their perhaps longer but no doubt less productive lives – see Key concepts box 3.1). Eventually there would come a point when natural selection would stop the feature from running too far away from its original function (peacocks can, for example, still fly).

> ### KEY TERM
>
> **Runaway selection** Ronald Fisher's hypothesis that a male characteristic that originally arose via natural selection can become highly elaborate due to the female-choice part of sexual selection.

DISCUSS AND DEBATE – SEXUAL SELECTION AND NATURAL SELECTION

Consider the following features of animals:
Large horns in bulls compared to cows
Nest building in birds
A male robin singing at dawn
A dog raises its leg up to urinate on a post
Excellent nighttime vision in cats.
Which of these features do you think was sexually selected and which was naturally selected? How did you arrive at these answers? What would you need to know in order to decide whether each trait came about via sexual or natural selection?

Figure 3.2 Peacock.
Source: Image copyright © Italay/Shutterstock.com.

Signaling theory – The cost of an honest display

The concept of sexual selection did not make much of an impact during Darwin's lifetime, or indeed for a century after he died (Fisher aside). During the last quarter of the 20th century, however, animal behaviorists began to return to this idea and, in addition to gathering evidence that supported it, developed it further. One well-regarded theory that advanced our understanding of sexual selection is known as **signaling theory**. Initially advanced by Amotz Zahavi (1975), signaling theory proposes that such ornaments are advertisements for genetic quality. It is as if the peacock is saying to peahens, "Look, I am of such fantastic genetic quality that I can even support this massive and costly tail." (Interestingly, because of this it is also known as the "handicap hypothesis," as in the burden of a golf handicap). Once this theory was proposed it was very quickly realized that such displays need to be "honest" (Waynforth, 2011). It is important that any signal is honest, otherwise it will tend to be ignored. One way of ensuring honesty is for signals to be costly so that only those animals who genuinely have the desired quality can afford to have them. (Incidentally, in our own species, it has been suggested that the reason expensive accessories such as Rolex watches and Louis Vuitton bags are so desirable is that they are honest signals of wealth, because only rich people can afford to buy them.) Honest displays (those that reflect true quality) are therefore necessary for one sex (usually males) to impress the other (usually females). When you think about it, if a peacock could produce a colorful tail that is not an honest signal, then it would not benefit peahens to choose him (since she is, after all, ultimately looking for good genes to pass on). In a sense, today we can expand the term signaling theory to "the theory of honest displays of high mate quality" (Hamilton & Zuk, 1982; Waynforth, 2011; Zahavi, 1975). But given this is a little long-winded – signaling theory will suffice. Returning to humans, some researchers have tried to explain risky behavior in young human males as their signaling their courage to (and thereby appearing attractive to) females (Milan, 2010; Crooks & Baur, 2013). Note that signaling theory provides a different angle on the relation between sexual selection and elaborate traits to Fisher's runaway theory since, rather than just being "sexy," ornate male characteristics signal good genes (good in the sense of being able to grow such an encumbrance and survive). Hence, whereas natural selection drives the sexes in the same direction – what's good for the goose is good for the gander – sexual selection can drive their characteristics apart.

KEY CONCEPTS BOX 3.1

Sexual selection and cannibalism in the redback spider

As we saw earlier on, males are sometimes regarded as the sex that lives fast and dies young. This can certainly be said of one species of spider. It has long been known that females of the Australian redback spider will, if they are able, consume their male counterparts as part of their copulatory behavior. Until the mid-1990s it was thought that such cannibalism was of benefit to the female (she gains a meal) but, for those that fail to escape their femme fatale, a major cost to the male (he loses his life). In 1996, however, Canadian animal behaviorist Maydianne Andrade (1996) discovered that being eaten by their mate is an adaptive trait for the males of this species and that it arose via sexual selection. As a part of his copulatory behavior, the male spider positions himself above the jaws of the much larger female. Then, for many males, during the transfer of sperm, the female eats him. Andrade found males that are consumed gain two reproductive advantages. First, cannibalized males copulate for longer and hence transfer more sperm than those that survive, and, second, once a female has cannibalized a male she is more likely to reject subsequent suitors. This is clear case of the male's genes benefiting from the behavior rather than the male himself. Moreover, this bizarre example demonstrates how, as Darwin was well aware, sexual selection can increase reproduction and at the same time reduce life expectancy. At least for males.

Why have sex? Comparing apomixes and parthenogenesis

Because most large organisms such as trees, birds and mammals make use of sex, we tend to perceive it as the normal way to reproduce. Most **microorganisms**, such as bacteria, however, generally reproduce asexually through cloning (making, barring mutations, genetically identical copies of themselves). Moreover, asexual cloning is also known in many **multicellular** organisms, including ringworms, sea anemones and even, on occasions, for some vertebrates, such as frogs, chickens and hammerhead sharks. Technically, in multicellular organisms, sexual and asexual reproduction are known as **apomixes** and **parthenogenesis,** respectively. In the case of apomixes, two **gametes** (sperm and egg), each of which has only half of the genetic complement, fuse to form a **zygote** (containing all of the genetic material) from which the organism will develop. In parthenogenetic species, however, an exact copy of all of a single parent's genes is passed onto each offspring via an unfertilized egg.

> ### KEY TERMS
>
> **Microorganism** Organisms too small to see with the naked eye, such as bacteria and protozoa.
>
> **Multicellular** An organism that is made up of more than a single cell.
>
> **Apomixes** The scientific term for sexual reproduction.
>
> **Parthenogenesis** Asexual reproduction by production of offspring from unfertilized eggs. Literally means "virgin birth."

The fact that both sexual and asexual reproduction exist in nature suggests that each has advantages for that species that is supported by natural selection. The fact that in some cases both exist in the same species (such as domestic fowl and frogs) means that there must have been times in our evolutionary past when the two methods of passing on our genes were in competition. This raises the questions what exactly are the cost and benefits of each method of reproduction and why did sex win out for most large multicellular organisms?

Why is sex so popular?

Sex evolved sometime around 3 billion years ago and is widespread in both animal and plant kingdoms today. This means that for hundreds of millions of years organisms got along quite nicely without it. In order to understand why sex evolved and has become so widespread, we need to consider both the strengths and the weaknesses of sex and asexual reproduction (see Table 3.1).

Looking at Table 3.1, the most important points to pull out are the fact that a sexually reproducing organism passes on only half of its genes to each of its offspring (see "Disadvantages of Sexual Reproduction") and the fact that sex is superior to asex in the **host–parasite arms race**. These points are surprisingly related, although in a roundabout way. To unpack the first point – if we consider fitness as the number of genes passed on to the next generation, then an animal

KEY TERM

Host–parasite arms race A relationship between host and parasite whereby improvements on one side of the equation lead to selective pressure for counterimprovements on the other side. This leads to cycles of adaptation and counteradaptation without either side ever "winning" the race.

Table 3.1 Advantages and disadvantages of sexual and asexual reproduction.

Advantages of asexual reproduction	Disadvantages of asexual reproduction
• Reproduction only requires one individual • No genetic changes (good if genes well suited to local environmental challenges) • All individuals can produce offspring • Twice as many genes passed on per offspring (compared to sex)	• Little or no variation in offspring • Due to lack of variation, offspring susceptible to same pathogens • Deleterious genetic mutations that arise will be passed onto all offspring

Advantages of sexual reproduction	Disadvantages of sexual reproduction
• Offspring have a great deal of variation (can be good if local competition is high or offspring disperse to different environments from parents) • Deleterious genetic mutations not passed on to all of offspring • Superior to asex in "host–parasite arms race"	• Two parents required to reproduce • Time and effort spent in seeking mates and in courtship • Combining genes from two parents can break up good gene combinations • Only half of genes passed on to each offspring from each parent (i.e., each parent discards half of its genes with each offspring produced)

producing two offspring has twice the fitness as one producing one offspring. Because a sexually reproducing organism contributes only half of its genes to an offspring and an asexual animal passes them all on, then an asexual animal will have twice the fitness per offspring as a sexually reproducing organism. The second point means that because the biggest killers of multicellular organisms (including our own species) are parasites, such as viruses and bacteria, then we may provide our offspring with variability in antiparasitic defenses (such as the immune system, which varies greatly between individuals) in order to fight back against the bugs. It is called an "arms race" because like the nuclear arms race of the Cold War, when the Soviet Union and the USA became looked in an escalation of these deadly weapons, as each side of the parasite–host equation comes up with evolved solutions the other side quickly evolves counteradaptations. Currently, it is this ability to vary the immune systems of our offspring that is considered to solve the problem of sex having to be twice as good as asex for organisms that suffer from parasite stress. (This is how the two points highlighted above are related.) Interestingly, the species that make use of asex are believed to suffer less from parasite load. This is often the case for microorganisms – hence, they more frequently manage without sex. It should be noted that, as with the nuclear arms race, a host–parasite arms race can in theory continue indefinitely since improvements on one side of the equation form the evolutionary pressure for improvements on the other side. This host–parasite arms race explanation of why we reproduce sexually has been termed the **Red Queen hypothesis** (Van Valen, 1973; Ridley, 1993) because in Lewis Carroll's *Alice through the Looking Glass* the Red Queen tells Alice that if you run at normal speed, you will stay in the same place – but if you want to move forward you have to run at least twice as fast! Hence to answer our question above – sex is so "popular" and widespread because it produces antiparasitic adaptations.

> **KEY TERM**
>
> **Red Queen hypothesis** The evolutionary hypothesis that proposes that organisms have to continue to evolve, not to improve, but to just remain as successful as their ancestors were, because other organisms (competitors, predators or parasites) are also evolving to gain an upper hand. The host–parasite arms race is a version of this.

Sexual selection and human intelligence?

In recent years sexual selection theory has been applied to help understand the evolution of certain aspects of human evolution and behavior. One aspect of behavior has been used to help illuminate is sex differences in human mate-choice criteria. We explore this relationship in Chapter 4. Surprisingly, it has also been used to help throw some light on the evolution of human language and intelligence – which we explore here.

KEY CONCEPTS BOX 3.2

Neoteny in men and women

Compared to our anthropoid relatives, humans retain a number of juvenile features into adulthood. In terms of behavior, as adults we retain a sense of playfulness and curiosity. Additionally, we are relatively hairless compared to the great apes (this is a juvenile feature in other primates). Interestingly, however, in terms of physical form, women are more **neotenous** than men. Despite maturing at an earlier age than men, women do not go on to develop skin as coarse as men. Neither do they develop a deepened voice, bony eye ridges or thyroid cartilage. These sex differences occur cross-culturally and are used by members of both sexes to distinguish gender visually (Jones & Hill, 1993). This raises the question as to why women are more neotenous than men. Because women end their period of fertility at a younger age than men, then it may have paid ancestral men to preferentially select women who showed signs of youth (which correlates with fertility – see Chapter 4). By pairing with younger females, males are more likely to produce more surviving offspring. This means that the reason women are the more **gracile** (slight and graceful) sex is probably down to sexual selection – but in this case can be traced back to male rather than to female choice.

Language and the "Sexy brain" hypothesis

In Chapter 2 we outlined two hypotheses that have been put forward to explain the evolution of human intelligence – the ecological and the social complexity hypotheses. Both of these hypotheses are based on natural selection. Since the turn of the century, however, a number of evolutionists have begun to suggest that our level of intellect did not arise from natural selection at all. In 2000, American psychologist Geoffrey Miller began to champion the notion that the driving force for the evolution of intelligence in general and language in particular is sexual selection. In his book *The Mating Mind: How Sexual Choice Shaped the Evolution of Human Nature,* Miller (2000) noted that people generally have a much larger vocabulary than is necessary for communication (a typical adult has a 20,000 **word family,** but the vast majority of communication can be accomplished with a 6,000 word family). In an attempt to explain this level of redundancy, Miller suggested that humans

---

KEY TERM

Word family The base form of a word and all of its derived forms, such as different tenses. As an example of this, think of the base word "run"— its family also includes "ran" "running" and "runners."

use their vocabulary in order to demonstrate their level of intelligence, and this, in turn, can be taken as a measure of "fitness" to a potential mate. This means that rather than developing physically sexy traits, such as large antlers or colored tail feathers, our ancestors developed a "sexy brain" in order to compete. His argument incorporates both male–male competition and female choice. According to Miller, those ancestral males that were best able to impress females with their communication abilities would have had increased mating opportunities. Note that such verbal fluency is also used to compete with other males. This raises one big question – if it's about males impressing females, then why do women also possess verbal prowess? Miller explains this as women using complex language in order to probe men and assess their quality. That is, they have to be able to use a complex vocabulary in order to determine just how intelligent a given man is.

In a sense we might think of this argument as the "chat up" theory to explain the evolution of complex thought and language. But Miller extends this hypothesis beyond just chatting someone up and convincing them you are clever enough to recombine their genes with. He also suggests it may extend beyond face-to-face flirtation as good communication may lead to an increase in social status and thereby make someone appear as a good potential mate. Hence, rising in status through the arts and sciences may be a route that is particularly attractive to males since it might then lead to an increase in mating opportunities. In support of his argument, Miller claims that males are 10 times more prolific in creating works of art and in publishing books. No doubt feminists would argue that in patriarchal society women have fewer opportunities to display art or publish books. To which Miller might reply – but you have first to explain why society is (or used to be) patriarchal? Interestingly, Miller also notes that males produce their most significant works of art, music and literature during their peak reproductive years (20–35).

Criticism and support for the mating mind

As you can imagine, the mating mind–sexy brain hypothesis has not been without its critics. Fitch, Hauser and Chomsky (2005), for example, consider that it is unlikely that sexual selection could have driven the evolution of communication of complex information about the world. They prefer instead to give the role to natural selection. But Miller also has a range of supporters, and a number of studies seem to suggest that sexual selection may well have been involved in the evolution of language and intelligence, with courtship in particular playing a major role. One study from the University of Nottingham found that following an imaginary romantic encounter with a young female, men then began to use more low-frequency words (in comparison with an encounter with an older female – Rosenberg & Tunney, 2008). Note that low-frequency words are the most rare and complex ones in a given language, and arguably they are the ones most likely to impress. In contrast, females in the study used *fewer* low-frequency words following an imaginary encounter with a young man. The authors of their study point out that vocabulary size correlates highly with measures of intelligence, suggesting the former is an honest signal of quality (see signaling theory, above). Overall, they consider that their findings provide support for Miller's hypothesis of males in particular using language to impress a potential mate and that this did arise from sexual selection.

DISCUSS AND DEBATE – USING LANGUAGE TO IMPRESS THE OPPOSITE SEX – OR NOT?

In the study by Rosenberg and Tunney (2008) briefly outlined above, while males increased the complexity of their language following an encounter with an imaginary young woman, females reduced the complexity of their language following a similar encounter with young male. Why might women be doing this? You should be able to think of more than one answer.

Can you think of any potential shortcomings of this study?

How might this study be extended?

Sexual selection and sex differences in human aggression

Cross-culturally, men are more likely to be involved in acts of direct aggression than women are. This is the case no matter which form of measurement is taken from grievous bodily harm to murder (Archer, 2004; Campbell, 2013b). Many social scientists explain this difference in terms of **social role theory**. That is that, due to feedback from others, such as parents, teachers, peers and the media, males and females internalize gender-stereotypical traits and then come to behave in gender-appropriate ways (Eagly, 1987). Hence, females will tend to develop

KEY TERM

Social role theory An approach that places pressures from society as the main determinants of behavior and internal states. Specifically in relation to gender roles, social role theorists generally reject biological/evolutionary influences.

KEY TERMS

Communal traits The capacity or tendency to act in friendly, unselfish and expressive manner. In social role theory females are socialized to show communal traits and males to show agentic ones.

Agentic traits The capacity or tendency to make choices and impose them on others.

Biosocial theory In relation to explanations of sex differences in behavior – the view that physical biologically influenced characteristics can have a secondary impact on the type of gender role a person gravitates towards. Note that the term has, however, been used in different ways by different social scientists.

communal traits (i.e., helpful, nurturant and kind) whereas males tend to develop **agentic traits** (i.e., self-reliant, dominant and aggressive). This means that as they develop, males become more likely to consider use of aggression a male-typical response under certain circumstances (such as when their pride is threatened). Females, in contrast, come to consider that aggressive acts are inappropriate for their gender and hence are far less likely to engage in direct aggressive responses (Eagly, 1987). Social learning theory has historically been considered a blank slate view of human nature – that is, we are not born with any particular behavioral propensities but rather society determines very much how we "turn out." In recent years, however, due to the rise of evolutionary approaches and a greater understanding of biological influences, a new form of social learning has emerged – the **biosocial theory** of development of gender roles (Wood & Eagly, 2002). This model of gender role development suggests that humans are born with biologically influenced physical sex differences – such as men being generally a bit larger and stronger than women. Then, as a consequence of these physical differences, we gravitate toward gender roles that are still heavily influenced by societal norms. In this model there are no direct genetically influenced psychological sex differences in the brain, but rather through feedback from our physical differences and from society we then gravitate to gender-appropriate roles. Sexual selection hence has no direct influence on psychological sex differences.

Evolutionary psychologists generally question both the traditional social role and the biosocial accounts of sex differences in aggressive response or, at the very least, suggest it is only a partial explanation. During the 21st century, evolutionists have begun to consider sexual selection theory as an explanation for this universal difference between the genders. One British social psychologist, who has examined the relative merits of sexual selection theory (SST) and both forms of social role theory (SRT) as explanations for human sex differences in behavior, is John Archer of the University of Central Lancashire. In order to determine which of these two major explanations best explains observed differences in rates of aggression between the sexes, Archer outlined a number of predictions that would arise from a social role/ biosocial explanation and from a sexual selection explanation. (These are outlined in Table 3.2.) Note from Table 3.2 that these two forms of explanation generally lead to different predictions. He then examined the large body of published studies in the field of sex differences in

Table 3.2 Predictions and findings regarding sex differences in aggression from sexual selection and social role theories.

Areas considered and predictions	Sexual selection theory (SST)	Social role/biosocial theory (SRT)	Findings
Magnitude & nature of sex difference	Largest differences will be in physical aggression, followed by direct verbal aggression. Indirect aggression shows either no difference or is more common for females. No difference in anger.	Magnitude will be modest. Larger difference for physical than for indirect (psychological) aggression. No difference in anger.	Data from published studies show large difference in physical aggression in male direction and difference in indirect aggression in female direction. Supports SST.
Development (Age trends in sex differences in physical aggression)	Early emergence of sex differences in direct aggression. Peak in damaging and risky competition during young adulthood (i.e., when competition for mates is highest).	Sex differences will start small and increase with age through childhood due to cumulative influence of socialization.	Studies show that sex differences begin early in life (based on observations studies, and mother's reports) and do not increase with socialization. Supports SST.
Mediators (Underlying causal factors, such as hormones, or situational factors, such as fear and other emotions related to aggression)	Mediators of aggression follow functional (evolutionary) principles (e.g., greater risk taking by males). Also greater fear of physical danger by females.	Mediators follow from characteristics of gender roles (e.g., fear of retaliation, guilt, anxiety associated with aggression, empathy). They arise from general learning mechanisms associated with gender roles.	Studies suggest two features that reduce aggression in females (fear and empathy) are reduced in women when administered testosterone. Suggests evolved hormonal differences are involved in sex difference. Provides limited support for SST.
Individual differences (Within sex, individual differences in levels of physical aggression)	SST suggests there will be a greater variability in males with regard to physical aggression depending on their emphasis on mating or on parental effort (i.e., males can take on "cad" or "dad" role). This should not hold for females as they provide a high level of parental effort.	SRT does not predict greater variability in physical aggression for men than for women.	Questionnaire based evidence suggests greater variability in levels of physical aggression for males. There is limited evidence that this are related to level of parental effort in males. Provides broad support for SST.

(Continued)

Table 3.2 (Continued)

Areas considered and predictions	Sexual selection theory (SST)	Social role/biosocial theory (SRT)	Findings
Variability in sex differences in aggression in response to environmental conditions	Variability across local conditions, cultures and nations is expected to reflect resources that are important for reproduction (e.g., lack of access to mates and the status and resources likely to increase physical aggression in males).	There will be variations in women's level of aggressiveness accompanying: (1) changes in role salience; (2) cross-cultural variations in women's relative emancipation; (3) changes over historical time in women's roles and status.	A number of studies found the level of violence is higher in men who have few economic resources or prospects that would aid access to females. No robust support for predictions 1, 2 or 3 in box immediately to left. Supports SST.

Source: Based on Archer (2009).

aggression to determine the degree to which findings supported either the social role or the sexual selection explanation. His findings are summarized in the right hand column of Table 3.2.

We can see from Table 3.2 that according to Archer, all five areas considered the findings (to various degrees) to support sexual selection theory as an explanation of sex differences in aggression rather than either form of social role theory. Due to this, Archer concluded that both the magnitude and the nature of sex differences in aggression can be better explained by sexual selection theory than by either form of social role theory. It is worth highlighting and explaining a number of the points outlined in the table. With regard to magnitude and nature of sex differences, Archer divides aggression into three descending levels:

1 Physical – actual violence towards another.
2 Direct verbal – aggressive language used directly towards another (often called *psychological aggression* since the aim is to hurt another's feelings).
3 Indirect – non-direct aggressive acts include the spread of negative gossip designed to harm another's reputation.

The evidence Archer uncovered suggested that males more commonly make use of Level 1 and females more commonly make use of Level 3. There is debate regarding which sex makes use of Level 2 most frequently.

With regard to development, social role (and biosocial) theory would suggest that as we are born psychologically gender neutral, differences in aggression will initially be minimal but will increase during development due to feedback as to what is appropriate behavior. The fact that such differences are there from a very early age suggests that sexual selection in the past has led to males being primed to be potentially more aggressive from birth.

In the case of mediators of aggression, these constitute a wide range of underlying causal factors, including situational and biological influences. The fact that Archer found that fear and empathy (which normally reduce aggression) were reduced when females were given testosterone is taken as evidence that this hormone normally has this role in males. Hence, in a roundabout way, this finding provides a degree of support for SST.

For individual difference (within sexes) the important point is that, according to SST, males can gravitate towards one of two extreme strategies – which we can broadly call "dads" and "cads." In the case of "dads," a male "settles down" with a female partner and spends his time helping to raise children (most frequently, but not exclusively, in a married relationship). For "cads" the strategy is to attempt to inseminate as many women as possible and provide little or no parental care. In reality, these are two extremes of a continuum of male-like behavior. A male can, for example, move from cad to dad and vice versa. Under SST, when a male gravitates towards a cad role, he is then more likely to come into direct competition with other males and is therefore more likely to be involved in aggressive confrontation then those who adopt a dad role. Since females are the sex that give birth, then during their evolutionary past they would not have had two potentially very different roles to choose from in the same way that males would have had. Put bluntly a female cannot leave a male "holding the baby." The fact that there is greater variability in levels of aggressive response in males than in females is taken as evidence that males can take on these two different roles, and this, in turn, is taken as supportive evidence for SST.

Finally, when we consider variability in sex differences in aggression in response to environmental conditions, what is suggested here is that males will show higher levels of aggression when their path to mates is thwarted by environmental conditions, such as a lack of resources. The fact that this is what is observed is taken as support for the SST by Archer. The three SRT derived predictions suggest that the level of aggression shown by women will change as does their role in society (e.g., they will be less subservient to men in societies where they have become more emancipated). The literature does not, in Archer's view, provide a great deal of support for these predictions.

Criticism and support for Archer's conclusions

Although Archer's arguments are based on a large body of evidence, not all those who have considered his stance agree with his conclusions. Two psychologists who were prominent in developing social role theory (and subsequently the biosocial theory) Eagly and Wood (2009) suggest that he has underestimated the flexibility that is built into human sex differences. They cite the fact that, under extreme conditions where male numbers have been reduced due to warfare, women have served successfully in fighting units in the army (including Eritrea's war of independence from Ethiopia). Archer does, however, allow for a fair degree of flexibility that is related to societal gender roles. This he considers is layered on top of differences that can be traced back to sexual selection. Others have been even more favorable of SST as a form of explanation than Archer. Kingsley Browne (2009), for example, suggests that the (limited) available evidence suggests that, when women have given combat roles, they tend to be less "trigger happy" than their male counterparts. Browne is also critical of the biosocial theory since, while it accepts physical differences as having arisen from evolutionary pressures, it fails to explain why those physical differences arose unless they evolved for competitive means.

DISCUSS AND DEBATE – SEXUAL SELECTION AND SEX DIFFERENCES IN AGGRESSION

Archer considers that sexual selection theory is a better explanation of sex differences in aggression – but importantly leaves the door open to social learning theory to explain some of the differences we see. Is it possible that both forms of explanation can be correct? If sexual selection is a better explanation for sex differences in aggression, might this have implications for policy makers (such as government bodies)? (Consider this question before reading the following paragraph below.)

The evidence that Archer presents suggests quite strongly that sexual selection theory may be a better explanation of sex differences in aggressive behavior than the more traditional social role/biosocial theory. Many of the academic debates in psychology have few serious consequences for the lives we live. This may not be so in this particular case. Given that a number of countries are currently considering changing the role of women in the armed forces and integrating them into frontline combat roles, this is one debate that transcends mere academic intrigue. If we can explain differences in the level of aggression shown by men and women as the outcome of social learning, then integration of women into such combat roles should be no more problematic than training men to fight. If, however, as Archer's review suggests, the origin of such sex differences arose from sexual selection pressures involving male–male competition, then we can expect that sex roles might be less malleable with regard to physical aggression. And, if that is the case, then a policy of integrating women into combat roles may be based on politically correct assumptions rather than on empirically correct findings.

Summary

- Characteristics such as gaudy male ornamentations are difficult to explain through natural selection and are considered to have evolved via sexual selection. Unlike natural selection, which selects for survival characteristics, sexual selection selects for characteristics that improve access to the opposite sex.
- Sexual selection can be divided into intrasexual selection, which is based on competition between members of one sex for access to the other sex, and intersexual selection, whereby one sex attempts to impress the other sex.
- Fisher's runaway selection hypothesis proposes that characteristics that were originally naturally selected (e.g., tail feather of the peacock aided flight) can, through sexual selection, become more elaborate to impress the opposite sex. That is, they runaway from their original function through intersexual selection.
- Signaling theory suggests that sexually selected ornamental features are both honest and costly. Hence they are genuine signals of quality in a potential mate.
- Sexual reproduction (apomixes) can be more costly than asexual reproduction (such as parthenogenesis, where an unfertilized eggs develops) because an organism throws away half of its genes when reproducing sexually. Sex, however, leads to variation in offspring, which may help them to overcome parasites. This explanation of why sex is so common is known as the *Red Queen hypothesis,* because it is based on a host–parasite evolutionary arms race whereby improvements on one side of the equation

lead to improvements on the other side (the Red Queen in *Alice through the Looking Glass* tells Alice she has to run very fast to stay in the same place).

- Darwin added a whole lexicon of terms related to his theory of sexual selection. The term *primary sexual characteristic* refers to those physical differences that have evolved for reproduction directly (such as the urinogenital system of each sex). *Secondary sexual characteristics* refers to characteristics that are not directly related to reproduction but rather evolved to aid same-sex competition and impress the opposite sex. *Sexual dimorphism* is a general term used to describe physical and behavioral differences between the sexes.

- Under the mating mind hypothesis, Geoffrey Miller has suggested that human language and intellect evolved via sexual selection as females used linguistic ability as a measure of "fitness" in males. According to Miller, females also evolved complex language and intellect in order to probe and assess the quality of potential partners.

- Social role theory has traditionally been used to explain sex differences in behavior, including levels of aggressive response. A new version of social role theory is called the *biosocial theory*. Under the biosocial theory, biological sex differences in physical attributes then contribute to differences in behavior. Males, for example, come to see themselves are larger and stronger than females and then gravitate toward different behavior patterns rather than having psychological differences that came about via sexual selection. Having reviewed the research literature, John Archer has concluded that sexual selection theory provides a more convincing explanation of sex differences in aggression than either form of social role theory.

FURTHER READING

Crooks, R. L., & Baur, K. (2013) *Our sexuality* (12th ed.). Belmont: Wadsworth.

Miller, G. F. (2000) *The mating mind: How sexual choice shaped the evolution of human nature*. London: Heinemann/Doubleday.

Ridley, M. (1996) *The origins of virtue*. London: Viking Press.

Toates, F. (2014) *How sexual desire works: The enigmatic urge*. Cambridge: Cambridge University Press.

Mate choice
The origin of human sexual preferences

4

What this chapter will teach you

- The evolutionary origin of sex differences.
- How sexual selection theory has been used to help explain differences between the sexes in their mate choice criteria.
- Criticism of applying sexual selection theory to help explain human mate choice behavior.

Mate choice psychology

Following on from Chapter 3, in this chapter we consider how **sexual selection** theory might be used to illuminate human mate choice predilections. In doing so we will consider developments since Darwin's time and what the research from evolutionary psychology can tell us about male and female reproductive behavior (and why these might differ at times).

Why choose?

As we look around the natural world we see that many organisms seem to be fussy about who they mate with, but why? Briefly, here is the explanation in a nutshell, and we shall develop this throughout the chapter and point out variations on this theme.

Recall that according to the **gene-centered** view of life outlined in Chapter 1, physical and psychological traits can evolve even when they disfavor the individual who possesses these traits (we discussed examples of self-sacrifice in Chapter 1). In a similar vein, mate choice preferences do not need to benefit the organism who expresses the preference, they benefit the genes that caused that preference to be expressed in first place.

Confused? Well, to start with, we have to assume that individuals vary in their overall genetic quality. Some individuals are healthy, fertile and strong; others, less so. Now, imagine that everyone mates with anyone without a second thought. If you are an organism with healthy genes and you mate with another organism that also has healthy genes, then on average your offspring will also tend to be healthy. Much healthier than if you inadvertently mate with someone who has unhealthy genes. Under such circumstances, sex is just a lottery: some you win, some you lose. Now imagine that by some combination of mutations an organism had the ability to determine the underlying genetic quality of a potential suitor. She eschews the advances of those individuals that she deems to have unhealthy genes and opts for the healthier ones. Her offspring are, on average, going to be healthier than those who are incapable of choosing – think of her advantage as being able to glimpse an opponent's hand in a game of poker. Healthy offspring are therefore more likely to survive to reproductive age and pass *their* genes on to the next generation. Furthermore, they are likely to inherit their mother's ability to determine genetic quality and pass that on to their offspring too. We can therefore see that the mate choice preferences could evolve using the same reasoning that we applied to the peppered moth in Chapter 1.

But surely this is fanciful, how can an organism determine the genetic quality of potential mates? The answer is through the effect that genes have on physical (and in some cases psychological) traits. We will discuss this more later, but as a taster we might assume that organisms that are good at resisting parasites, for example, will show this in being relatively unblemished compared to their parasite-laden conspecifics: to a very real extent, we wear our genetic quality on our skin.

So choice is something that matters and genes that enable their organisms to choose wisely are rewarded by being represented in future generations, those that don't, aren't. But, as we shall see, making good and bad choices may have different effects on different organisms and different sexes. Before that, it is worth spending a few minutes thinking about sex and the different sexes that engage in it.

DISCUSS AND DEBATE, SIGNALS OF GENETIC QUALITY

Before we consider the differences in what males and females consider to be attractive in the opposite sex, it is worth also considering the similarities. Make a list of the things that you think both males and females consider to be attractive traits. These should be physical and psychological.

The evolution of sex differences

Sex without sexes

Despite its racy image, sex is really quite unsexy as it simply describes a particular process of genetic mixing that increases the genetic variation of offspring, perhaps to help resist fast-mutating parasites (see host–parasite arms race– Chapter 3). Whatever the advantage of sex is, as we saw in Chapter 3, it does seem to have some advantage as it is the reproductive method of choice for the vast majority of **multicellular** organisms.

Sex was probably even less sexy in the beginning as it wasn't even between males and females. Instead, the first sex was between single-celled eukaryotes (cells with a nucleus) who produced gametes that were almost indistinguishable from one another – no sperms and eggs here. This form of sex is known as **isogamy** and is still practiced in organisms such as mosses. In this early form of sex there were no males and no females, just two individuals who contributed the same amount to the offspring, but as organisms became more complex it was necessary for the developing offspring to be provided with rather more than just genetic material: they needed food and nutrition as well. It may have been possible for both parents to divvy up the requirement of the offspring's nutritional requirements equally, but instead, asymmetries evolved so that one parent supplied only genetic material and the other was burdened with supplying everything else, nowadays we call the former males and the latter females. As you can see from

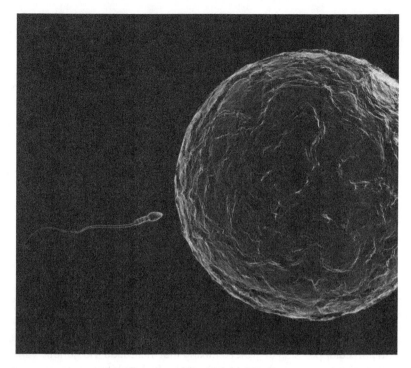

Figure 4.1 Picture of sperm with ovum, showing relative size.
Source: Image copyright © James Steidl/Shutterstock.com.

Figure 4.1, the female ovum really is a massive structure compared to the tiny, but highly **motile,** sperm.

The females of many extent species cried enough at this level of male–female inequality (most female fish, for example, drew a line under any greater contribution, and some – for example, the seahorses – demanded a higher contribution from their partners, and so it is male rather than female seahorses that become "pregnant"), but the females of many other species continued down this high-investment route. Fertilization, hitherto conducted in the outside world – females would lay eggs and males would fertilize them – was brought in-house in many birds, reptiles and mammals. So now females hosted the developing embryo, for a time at least, expelling them as large eggs, and then – again, in most cases – taking up the lion's share of brooding when compared to their male counterparts. Other females took it even further and gave birth to their offspring full term. These are the so-called **viviparous** animals, often misdescribed as giving birth to "live young." Unless something very wrong has happened, all animals produce live young; it is better described as the offspring being born as a smaller version of the adult rather than as an egg.

In mammals the female contribution reaches an extreme. Not only do mammals give birth to full-term offspring (with the exception of **monotremes,** such as the echidna and the duck billed platypus, which both lay eggs) they have also developed special organs specifically to feed their offspring once they have been born: mammary glands, by which they manufacture milk.

If one were to take a historical view of the evolution of sex from its most ancient times to the more recently evolved species, one might note a worrying trend: things have become less equal, with mothers, on average, investing more and more in their offspring relative to fathers. There are some exceptions to this – seahorses, as mentioned above, and jacanas, which we will meet later. In both these species the males invest a large amount in offspring – in some cases, more than the female.

What about the guys? Mating systems

Across the animal kingdom, males show different levels of male parental investment (MPI – see Chapter 5), with males of some species investing highly in offspring either directly, by feeding or otherwise engaging in childcare, or indirectly, by provisioning the mother. It should be noted, and as we suggested previously, that across species, females also show greater or lower levels of investment as well, although in the majority of cases for any given species the female invests more than the male. We can capture the relative male investment by investigating mating systems – the most common four of which are described below.

- **Monogamy** is where one male and one female mate exclusively with one another and stay together for a period of time (the term **pair-bonding** is also used to describe this kind of relationship). Monogamy, or pair-bonding, is particularly widespread in birds, with more than 90% of species subscribing to this mating strategy

(Lack, 1968). Males in such relationships frequently devote a degree of time and effort to raising their offspring, particularly in provisioning of food.

- **Polygyny** is where a male has (usually exclusive) mating access to more than one female. Quite often this involves one male having a **harem** of females. Polygyny is particularly common in mammals, with species such as elephant seals, deer and gorillas engaging in this mating system. Males usually play little or no role in rearing the offspring, their contribution to parenting being little other than "a few minutes of sex and a teaspoon of semen," as Steven Pinker (1997, p. 468) put it.
- **Polyandry** is where a single female has access to a number of males. This is a relatively rare mating system in the animal kingdom.
- **Promiscuity** is where there is no long-term commitment by either males or females to each other following mating. Across the animal kingdom, this is the most common mating system of all and in most cases neither the mother nor the father play any role in the rearing of offspring once fertilization has occurred. Examples in primates include chimpanzees and bonobos who, relative to humans, are much lower in MPI.

Sex differences in mate choice: Are men from Mars and women from Venus?

The title of this section refers to a book by John Gray (1992), which suggests that men and women differ in the way that they evaluate relationships. The term has since become used to describe the fact that men and women are fundamentally different. Unfortunately, evolutionary psychology has become associated with making similar claims. Implying, for example, that men are aggressive, competitive, concerned with dominance and largely interested in promiscuous sex with multiple partners but are somehow trapped into relationships with women who are nurturing, caring and empathetic (see Chapter 3). Furthermore, because this is evolutionary psychology, it is often assumed that these differences are purely genetic and therefore and in some ways "natural." Perhaps, unsurprisingly, these findings have been met with a storm of protest by those who believe that they are merely reinforcing old stereotypes and standing in the way of attempts to promote sexual equality (we discuss this more in Chapter 11). So what is going on here? Does evolutionary psychology really suggest that men and women are fundamentally different? And does this undermine the notion of sexual equality? In the next section we address the first of these questions, and to reveal the punchline, the answer is a qualified yes. There are good reasons why we should expect there to be differences between men and women in decisions that are related to reproduction and good evidence to that these differences exist. But these differences are frequently overstated in popular literature. Furthermore, and as we discuss later with reference to lonely hearts advertisements, to a very large extent, men and women want very similar things in a partner.

In answer to the second question, we should not consider that a moral argument can be used as evidence for a factual argument – a mistake known as the **naturalistic fallacy**. Nor should we deny the existence of something merely because it doesn't seem desirable from a moral point of view – a mistake known as the **moralistic fallacy**.

From an evolutionary perspective, there are at least four reasons why we can predict that there may be differences between men and women when it comes to reproductive behavior. Such differences can be thought of as asymmetries between the sexes and include the potential number of offspring produced, length of period of fertility, the cost of reproduction and the possibility of cuckoldry. In order to understand why the sexes may differ in their mate choice behavior, it is necessary to have some understanding of each of these differences. Hence, we briefly examine each in turn.

> **KEY TERM**
>
> **Naturalistic fallacy** The misguided attempt to derive a moral "ought" argument from an empirical "is" argument. An example might be that men on average are paid more than women, hence it must be morally correct that men should be paid more than women. The **Moralistic fallacy** is the opposite – the attempt to derive an empirical "is" argument from a moral "ought" argument. For example, it is good for both males and females to be equally interested in children, so we deny any evidence that this is not the case (and overemphasize the evidence that it is the case).

1. Asymmetry in potential number of offspring produced by males and females

As is the case for all vertebrates, human males produce enormous quantities of sperm, whereas human females have a limited number of eggs (technically, unfertilized **ova**). While a teenage girl beginning her periods has 400 ova that will be released over her fertile years, a young, sexually mature male produces new sperm at a rate of over 10 million every hour. Moreover, whereas a female can only become pregnant once every 9 months (in reality, due to a number of factors, interbirth gaps are normally a fair bit longer than a year), a male can, in theory, make a different female pregnant every day. Hence, beginning with this difference in numbers, we can predict that females will be more careful with their gametes than males. This, however, is only the beginning of how the sexes differ in reproductive matters.

2. Length of period of fertility

A human female is typically fertile between the ages of 13 and 50 (although the chances of becoming pregnant after the age of 40 begin to diminish quite rapidly – see Figure 4.2). Males, in contrast, can produce viable sperm from the age of around 13 up until old age (the current record for fathering a child is 96 year old Ramajit Raghav, from India). Hence, given women's relatively short period of fertility, it has been argued that there will have been stronger selective pressure on ancestral men to identify and court females who show signs of youthfulness than for women to court young men.

3. Asymmetries in parental investment

As we have seen, Darwin noticed that sexual selection led to attractive males who competed with each other for the attention of the females of their species. He was unable, however, to explain why it was this way around. Why should males in the animal kingdom be the sexy and aggressive ones rather than the other way around? In 1972, a young evolutionist by the name of Robert Trivers solved Darwin's conundrum. Trivers realized that intersexual selection (i.e., male attempts to impress and female reluctance to be impressed) is the result of an asymmetry in the cost of reproduction for males and females. Because, as we saw earlier, females generally put a great deal more in the way of resources into reproduction than males (think of gestation, lactation and parental care in mammals and the internal development and brooding of large

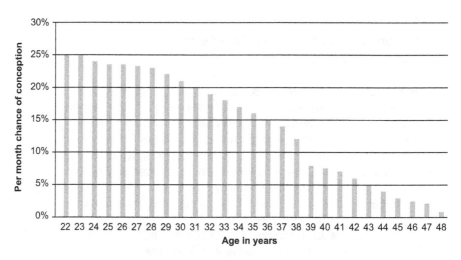

Figure 4.2 The average decreasing chances of conception as a women ages.
Source: Dunson, Colombo, & Baird (2002).

eggs in birds), males are apt to compete for those female resources. In a sense, males are competing for females because the latter already bring a lot to the table, whereas females are looking for males that in some way bring something to the table to help even out the effort they have to put in. Trivers called the amount of effort that males and females put into the production of offspring **parental investment**. He also suggested that it is this asymmetry between the sexes, with females investing a great deal more, which led to males attempting to be attractive to females and to compete with each other. We will return to parental investment later on and again in Chapter 5, but for now suffice to say that this asymmetry between the sexes has been used by evolutionists to help explain differences in behavior between them.

The great triumph of parental investment theory is that it can explain the exceptions. Although the finding that males compete and females choose is true for the majority of vertebrates, in some species it is the other way around. Consider the lily trotters. In these North American shore birds (also known as jacanas), the female is larger and more aggressive than the male, and it is the female who competes for other males. Parental investment theory can explain this apparent contradiction because it is the male who invests more in offspring; he is solely responsible for incubating the eggs and rearing the hatchlings. It is therefore the male of this species who has more to lose by making a poor choice of mate and is therefore more choosy (Emlen & Wrege, 2004).

Figure 4.3 A lily trotter, or jacana.

Source: John Hill [CC BY 3.0 (http://creativecommons.org/licenses/by-sa/3.0)], via Wikimedia Commons.

4. Cuckoldry

If men have an advantage over women in that they can, in theory, create a baby far more often than, say, once a year (and that they have a far larger number of gametes and a longer period of fertility), women, for their part, have what might be considered their own built-in advantage in the game of reproduction. Whereas when a female becomes pregnant she knows it is her child, her male partner cannot be certain it is his. Given that a man can end up bringing up another man's child (i.e., only males can be **cuckolded**) evolutionists have argued that there will have been selective pressure on males to develop measures to avoid cuckoldry. It is difficult to provide an accurate figure on the percentage of men that are deceived by their partners into raising another man's child, but experts in this area place the figure at around 10% (Bellis & Baker, 1990; Gaulin, McBurney & Brakeman-Wartell, 1997). Some evolutionary psychologists have even suggested that the observation that, cross-culturally, males attempt to control the behavior of females is directly related to the possibility of their being cuckolded (Daly & Wilson, 1983).

Predictions based on these differences

From these differences/asymmetries between the sexes, we can make a number of predictions about how the sexes may differ in their mate choice behavior. Given Points 1 and 2 (differences in potential number of offspring produced and in period of fertility), we can predict that males will seek signs of youthfulness in females, since these are indications of **reproductive value** in women. Reproductive value is a measure of potential number of future offspring. Hence, a girl of, say, 16 will have a higher reproductive value than a woman of 36. This means that what men perceive as attractiveness in women will be more closely related to signs of youthfulness (provided they are of reproductive age). We can also predict that, given Point 3 (the greater investment in reproduction by females), women will be choosy about male partners and that they should find signs of wealth, status and industriousness to be highly valued. This is suggested since older men will often have accumulated greater wealth and social status. Hence, we can also predict that men will be attracted to younger women (and increasingly so, as they themselves age) and women will be attracted to older men (since they are more likely to have gained wealth and status). Taking Points 1, 2 and 4 together, this means that the **market**

KEY TERMS

Reproductive value Anticipated number of future offspring production as a function of age. First used by Ronald Fisher (1930).

Market value In evolutionary psychology, ultimately an estimate of how valuable a partner is in terms of production of future offspring. In more proximate terms, how attractive a proposition a potential mate is in terms of characteristics including physical attractiveness, resources and social standing.

value of men and women will be partly based on different criteria, with signs of youthful attractiveness being highly rated for women and signs of resources being more highly rated for men. Finally, we can predict that men will tend to use countermeasures to avoid Point 4 – cuckoldry.

KEY CONCEPTS BOX 4.1

Sexy women and choosy men – A result of high MPI?

One result of sexual selection is that throughout the animal kingdom we encounter sexy males and choosy females. If this is the case then we might ask, Why aren't human males sexy, and why, when it comes to long-term partners, are men actually relatively choosy? Moreover, why is there general cross-cultural consensus that women are the attractive members of our species? One suggested explanation for all of these factors is that it is due to something called **high male parental investment (MPI)**. As we have seen, parental investment consists of how much time and effort each parent expends on the production of offspring (Trivers, 1972). Evolutionary psychologists consider that because human infants are born so immature, it paid males to provision the nurturing mother to increase survival rates of the offspring (unlike humans, 99% of male mammals provide no care for offspring). Once males began to provide parental care, they then become more choosy about whom they gave that support to, so the argument goes. Hence, men, being high in MPI, show a degree of selectivity when considering a long- term relationship with women (Buss, 2014; Waynforth, 2011). This choosiness is considered to be based on signals of fertility in females (especially given this period is more restricted than in males). So this high level of MPI for human males coupled with females need to signal fertility may also help to explain why, unlike other females, women have evolved sexually selected ornaments to attract the opposite sex and keep him around (permanently swollen breasts and a low waist-to-hip ratio, see below, that between them lead to an "hourglass" figure). These constant signs of fertility (other female mammals only show signs of fertility at times when they are about to release an egg) lead males to find them attractive throughout their estrus cycle. Hence, high MPI and a limited period of fertility in females led our species to show a degree of reversal of the predominant pattern of male sexiness and lack of choosiness compared to females. Why men aren't sexy is still an area of debate – but one suggestion is that rather than develop physical sexiness, such as the red face of the mandrill, human males display intellectual sexiness (Miller, 2000; see Chapter 3).

DISCUSS AND DEBATE – HOW MIGHT MALES AND FEMALES DIFFER IN THE FEATURES THEY FAVOR IN THE OPPOSITE SEX?

Before reading the next section, based on sexual selection theory, what differences would you predict men and women will show with regard to the features they would favor in a mate? Do you think these would fit in with our every-day view of how the sexes differ?

What do men and women want – The findings from evolutionary psychology

Over the last 25 years, a number of evolutionists have attempted to determine whether our criteria for choosing a mate do actually differ between the sexes along the lines that sexual selection theory would suggest. Such studies include large-scale questionnaires rating desirable mate characteristics, analyses of lonely hearts ads and consideration of physical attributes, such as hip-to-waist ratios and degree of robust masculinity of features. In many ways men and women want the same thing. They generally want someone who is understanding, kind, prepared to invest in them and physically healthy. This last factor is signaled by a number of "honest" physical features, such as facial and bodily symmetry. (Research shows that both males and females generally prefer symmetrical faces, and this has been shown to indicate a good immune system; Swami & Salem, 2011.) But there have also been areas where the sexes differ. We consider the most prominent findings in relation to sex differences from a number of studies below.

Cross-cultural similarities in mate choice preferences – the work of David Buss

One way of determining whether the various predictions arising from sexual selection theory regarding mate choice preferences really are fulfilled is simply to ask people. One of the first people to do this within an evolutionary framework was University of Texas psychologist David Buss. In the late 1980s, Buss and his colleagues collected questionnaire data from 37 different cultures around the world, providing a sample of just over 10,000 respondents. The questionnaires asked participants to rate on a 0- to 3-point scale various attributes in a long-term romantic partner, ranging from *good looks* to *good financial prospects* and *pleasing disposition*. The fact that he chose such a wide range of cultures is important because if respondents have similar preferences, this might be taken to suggest that just below the surface of cultural diversity there lurk common evolutionarily influenced tendencies. Alternatively, if there are great differences between cultures, then this might be taken to suggest that arbitrary cultural practices play a major role in the development of these mate choice preferences.

Interestingly, Buss uncovered evidence both of cultural diversity in preferences and of **universality** (i.e., the same pattern cross-culturally). Table 4.1 highlights the findings of Buss in relation to the predictions of sex differences based on sexual selection as presented earlier.

KEY TERM

Universality In evolutionary psychology, *universality* means that despite cultural difference, there are facets of human nature that are found in all cultures and are related to our evolutionary past.

In summary, Table 4.1 provides quite good support for the notion that physical attractiveness in a woman is more important to men than the reverse. It also provided strong support for the notion that men prefer to marry women who are younger than themselves. Both of these are conducive with the prediction that due to women's shorter period of fertility, men have evolved to choose women that signal fertility and hence high reproductive value. For their part, it is clear that women do favor men who are older and able to provide resources (that might be used to boost offspring survival) since they rate ambition and industriousness more highly in a partner than men do. (Incidentally, Buss et al. also found that women rate good financial prospects more highly then men do.) We should not ignore the fact, however, that there are quite large cultural differences for some of these preferences, suggesting that cultural practices are also of importance in determining what is important to each sex.

This study from 1989 was only the beginning for David Buss, as he has continued to study the relationship between Darwin's sexual selection and human mate choice preferences since then. His research includes studies of conflict between the sexes, sexual jealousy and how the sexes may differ in characteristics of long- and short-term relationships all drawing on evolutionary theory. It would be beyond the scope of this book to examine all of these in detail, but Key concepts box 4.2 outlines one of his recent studies on mate choice preferences and how they differ between the sexes when considering long- and short-term relationships.

Table 4.1 Mate choice predictions and findings.

Characteristic and prediction	Findings	Level of cross-cultural variability
Age preference. Males will prefer younger wives; females will prefer older males	Males preferred on average a marriage partner to be 2.66 years younger and females preferred on average that their marriage partners to be 3.42 years older	High cultural variability with the highest age difference being in Nigeria and Zambia, with males preferring females being 6.45 and 7.38 years younger, respectively. While this suggests variability, the general preference for younger females was found in every culture surveyed
Physical attractiveness. Males will regard attractiveness more highly than females in a marriage partner	In all 37 cultures, males rated physical attractiveness more highly than women (in 34 of these the difference was significant)	Low variability perhaps suggesting a general evolved phenomenon
Ambition and industriousness. Women value ambition and industriousness in men more highly than men do in women	In 34 out of the 37 cultures (92%), women value ambition and industriousness more highly than men – providing reasonably good support for the prediction	Moderate – but it should be noted that in three cultures – Colombian, Spanish and South African Zulu – the men rated this more highly than women. This may be taken to suggest that cultural practices (e.g., in Zulu society women are the house builders!) are not without importance here
Chastity. Men will value chastity in women (i.e., no previous experience in sexual intercourse) higher than women will in men. This is proposed as a strategy to increase certainty of paternity (i.e., only men can be cuckolded)	62% of cultures demonstrated a sex difference here conducive with the prediction. The remaining 38% show no significant sex difference either way. This suggests weak support for the evolutionary hypothesis that males will find chastity more important	Quite high, suggesting perhaps that cross-cultural religious considerations and availability of contraception may play important roles here

Source: Based on Buss (1989).

Lonely hearts advertisements

If sexual selection selects features that help individuals both to compete with their own sex and to attract the opposite sex, then we can expect members of our own species to emphasize such features when looking for a romantic partner. We can also expect members of each sex to make explicit the sort of features they would favor in the opposite sex. One way in which evolutionary psychologists have determined the extent to which people do emphasize and seek such sexually selected features is by extracting data from "lonely hearts" advertisements in newspapers and magazines. It is worth noting at this point that unlike the questionnaire used by Buss, extracting data from these ads avoids the potential problem of respondents complying with the experimenters' expectations (often a problem in questionnaire-based psychological studies). It is also worth pointing out that while Buss's study asked what people would ideally like to see in a mate, personal ads placed to request a mate have to take into account the market value of the advertiser. Hence, while the descriptions offered are likely to be positively framed, most people, having a notion of their own market value, will adjust their demands appropriately.

An example of such a study was published by Bogus Pawlowski and Robin Dunbar in 1999. As can be seen in Figure 4.4, women sought men that are older than themselves (dark bars) whereas men sought women who were younger than themselves (shaded bars). This fits in well with Buss's study described above. As Pawlowski and Dunbar (1999) stated, "Age is a criterion that males use in judging the mate value of prospective female partners, because it correlates with fertility" (p. 53). In addition to this finding, Pawlowski and Dunbar also discovered that the individuals most likely to omit or disguise their age were women between 35 and 50 (i.e., women who were coming toward the end of their fertile years).

Subsequent research by Dunbar and others has established that in addition to this age difference, women tend to advertise signals of physical attractiveness and seek financial security, whereas men tend

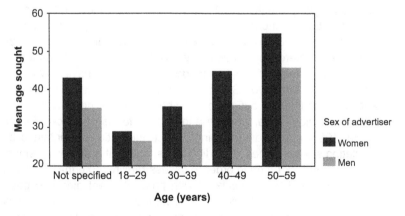

Figure 4.4 Mean age sought for a romantic partner by male and female advertisers in lonely hearts columns (also, age range of advertiser provided).

Source: Pawlowski and Dunbar (1999).

to offer financial security and request physically attractive women. Moreover, women regarded as physically attractive demand higher levels of resources in a potential partner (see, for example, Gurven & Kaplan, 2011; Pawlowsky & Dunbar, 1999; von Rueden, Waynforth, 2011; Waynforth & Dunbar, 1995). Hence, the findings from analyses of lonely hearts ads tend to back up those of Buss.

DISCUSS AND DEBATE – SEXUAL AND ASEXUAL REPRODUCTION

Some species (e.g., aphids) can use both **apomixes** and **parthenogenesis** under different circumstances. What predictions would you make about the circumstances under which members of a species would shift from asexual to sexual reproduction?

Waist-to-hip ratios

While questionnaires and lonely hearts ads draw on rating scales and personal requests for a romantic partner, Devendra Singh, of the University of Texas, has taken a more specific approach. Bearing in the mind the notion that sexual selection should lead to the evolution of **honest signals** (see above), Singh has proposed that one such signal is a low "waist-to-hip" ratio in women. This means, when the circumference of the waist is small relative to the hips, this is an honest signal of fertility and one that men universally find attractive (Singh, 1993a, 1993b; Singh & Singh, 2011). Typically, the waist circumference is divided by the hip circumference, and both high fertility and male optimum ratings converge on the figure of .7 (i.e., the waist is 70% the size of the hips). Interestingly, the winners of the Miss America pageants and the centerfolds of *Playboy* magazine invariably have a waist-to-hip ratio of .7. Moreover, Singh also claims that women with this low ratio are less likely to suffer from a whole range of health disorders, such as cardiovascular disease and various forms of cancer (Singh, 1993a, 1993b). According to Singh, since this ratio is maintained by circulating sex hormones (in particular, estrogen), it is an honest signal that men can pick up on in order to assess fertility (note that the more tubular ratio of 1.0 is associated with preadolescent girls and postmenopausal women). Hence, the argument goes, men who found this low waist-to-hip ratio (an honest signal of health and fertility) appealing would have left more surviving offspring than those who preferred a higher ratio.

KEY CONCEPTS BOX 4.2

The relative importance of faces and bodies to a partner

Imagine you have been invited on a "blind date," with a difference. Rather than having no idea what your date looks like, you are provided with a photo of him/her. But there is a catch. The photo is beneath the stick figure drawn on the two boxes below – and you can only remove one of the boxes to reveal either the face or the (clothed) body of your blind date. The question is, which box do you remove – would you rather see the body of your date or the face?

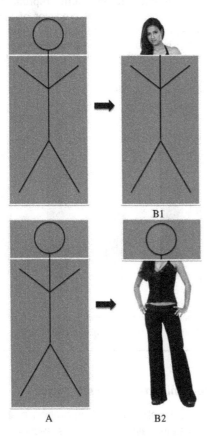

Figure 4.5a The relative importance of faces and bodies to a partner.

Source: Confer Perilloux and Buss (2010). Reproduced with permission.

This, in effect, was the question David Buss and his coworkers Jamie Confer and Carin Perilloux posed 375 students. So which box did they remove – face or body? What they found was by no means straightforward but really quite interesting. The majority of males and females preferred to remove the box covering the face (61% and 69%, respectively). In addition to this general finding, however, the researchers found that the relative importance of face and body was dependent on whether the students were asked to consider the potential mate for a one-night stand or for a long-term relationship. When the one-night stand option was thrown into the equation, men assigned greater relative importance to the body than to the face. In contrast, women did not show this shift, considering the face more important for both short- and long-term relationships (see Figure 4.5b). The researchers put this sex difference down to the fact that while a women's face provides accurate cues to long-term reproductive value (facial symmetry, for example, is a good predictor of this), her body may provide a greater indication of current fertility. They suggest that "men, but not women, have a condition-dependent adaptive proclivity to prioritize facial cues in long-term mating contexts, but shift their priorities towards bodily cues in short-term mating contexts" (Confer, Perilloux & Buss, 2010, p. 348).

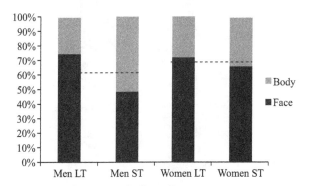

Figure 4.5b The relative importance of faces and bodies to a partner.

Source: Confer, Perilloux and Buss, (2010). Reproduced with permission.

Dads, cads and mixed-mating strategies

The work of Buss, Dunbar and Singh on how mate preferences differ between the sexes in ways that sexual selection theory would predict has helped in our understanding of this complex area. Over the last

20 years such studies have burgeoned, and today many aspects of the relationship between evolutionary theory and which features are important in mate choice decisions has become a major area of research. David Perrett and his colleagues at the University of St. Andrews have, for example, uncovered evidence that while men prefer facial features that correlate with youth and fertility, such as clear skin, large eyes and lips, a small nose and chin, women prefer signs of maturity in male faces, such as prominent eyebrow ridges and a well-developed jaw (Perrett et al., 1998). Or rather, women prefer these features during the fertile phase of their estrus cycle. Interestingly, there is evidence that they shift their preference toward less robust features in a man at other points in the cycle. One interpretation of this curious finding is that women may at times play a **mixed-mating strategy**. This means that when they are most likely to conceive, women shift their preference to cues that signal high levels of testosterone in men (robust masculine features correlate with high levels of this hormone), but during periods when they are less likely conceive, they shift preference to less dominant men who will make for better fathers. Hence, the mixed-mating strategy might reap benefits of good genes from one man and good care from another – a "cads" and "dads" mixture. This raises the question what is in it for the cuckolded "dad?" One idea is that given that only a limited number of males are able to offer good genes, perhaps a market. in covert matings has been produced (Dunbar, Barrett, & Lycett, 2005). Hence, a market of caddish behavior may be created where "dads" are prepared to put up with the odd child being a possible non-relative, provided that the majority of children are. This means, of course, that the behavior of each sex feeds back on the behavior (and preferences) of the other.

> ## KEY TERM
>
> **Mixed-mating strategy** The notion that some women may engage in supposedly monogamous pair-bonding with a caring male and simultaneously have extramarital affairs with other males with perhaps better genes in order to improve the viability of their offspring.

Based on the findings discussed above, evolutionary psychologists today suggest that each sex has inherited psychological adaptations to find attractive those features that would have boosted reproductive success in ancient past (Buss, 2014).

You may not be entirely happy or convinced by this and other evolutionary-based arguments for our mating proclivities. If that is the case, then you are not alone. A number of critics have suggested alternate explanations for the findings of evolutionary psychologists in this area. David Buller, in particular, has criticized the work of David Buss's large-scale study on mate choice preferences (see Key concepts box 4.3).

KEY CONCEPTS BOX 4.3

Criticisms of the concept of evolved preferences in human mate choice

The biological philosopher David Buller has been critical of much of evolutionary psychology to date. One of the major themes of his 2005 book *Adapting Minds: Evolutionary Psychology and the Persistent Quest for Human Nature* (Buller, 2005a) was a criticism of the notion of evolved mating preferences and, in particular, the findings of David Buss. Buller suggests that rather than men being adapted to find youthful women attractive and women being adapted to seek out older men, the age difference in Buss's cross-cultural study can be explained in a different and arguably simpler way. Rather than this age difference being due to evolved universal mating preferences, Buller contends that men and women are simply attracted to those who are of similar sexual maturity. Hence, given that the average young woman reaches sexual maturity 2 or 3 years prior to the average young man, romantic pairings reflect similar levels of maturity. Buller contends that this is superior explanation of Buss's findings and thereby dismisses Buss's explanation. Buller's view is certainly one that should be considered as a serious alternative to Buss's. Unfortunately for Buller, his hypothesis has two serious holes in it. First, Buller seems to think that because he has come up with a viable alternative, then this is reason enough to dismiss Buss's hypothesis. This is a rather strange view of how science works as in many fields competing hypotheses are brought to bear on the same data with new empirical findings regularly being reported in order to determine which is the more fitting explanation. Second, previously published independent research refutes Buller's hypothesis in favor of Buss. Kenrick, Gabrielidis, Keefe, & Cornelius (1996) demonstrated that males in their mid-teens favor older females as maximally desirable and it is only when they grow older that they shift their preference to women who are younger than themselves. If Buller's hypothesis is correct, then those young men should prefer girls who are younger than themselves (given his arguments about girls maturing at an earlier age than boys), but their predilection for older women, in peak years of fertility, goes against this notion. Moreover, Buller's suggestion does not fit in with the finding from lonely hearts ads that as they age, men prefer progressively younger women (Kenrick & Keefe, 1992). Both findings are more conducive to Buss's evolutionary hypothesis for mate choice preferences.

Cultural variability with regard to sex differences in mate choice preferences

While most evolutionary psychologists consider that these studies have uncovered a great deal about the relationship between sexual selection and human mate choice preferences, like David Buller, many social scientists remain to be convinced. It has been pointed out that, even in Buss's study, there were quite large cultural differences (Eagly & Wood, 1999). Another example of this cultural variability is the finding that while British women are three times more likely than British men to seek financial security (Greenless & McGrew, 1994), for Japanese women the figure is 31 times more likely (Oda, 2001). This is a clear cultural difference. Note, however, that in both countries, as in Buss's study, women seek this in a man far more often than the reverse. Much of the cultural variation appears to be in the degree of differences between the sexes rather than whether or not such differences exist.

It is clear that cultural factors also play a role in many of these sex differences. Eagly and Wood (1999; see also Eagly & Wood, 2011) have, for example, pointed out that the preference for men with resources diminishes somewhat in countries where women have access to financial rewards. This does not disprove Buss's hypothesis of an adapted response because, like evolutionary psychologists in general, he considers human mate choice responses to be partially contingent on local ecological and cultural pressures. To state that responses are contingent does not mean that they are infinitely flexible but rather that they show a degree of plasticity that is, in part, molded by these prevailing pressures. If they were entirely flexible then we would not expect to see preferences for, say, age differences to be universal. Just how flexible and culturally dependent these factors are is, however, an area of currently unresolved debate (Eagly & Wood, 2011, 2013).

Do evolved preferences lead to improved reproductive success?

We also have to ask to what extent do these factors uncovered by Buss, Dunbar, Singh and Perrett convert into real reproductive success? One field study that supports the work of Buss, in particular, is a recent demonstration by von Rueden et al. (2011) that dominant and prestigious men in a forager society (the Tsimane of Bolivia) do have more surviving offspring. This increase in inclusive fitness is due to a number of different factors such as higher fertility, marrying younger women and a greater number of extramarital affairs. Such a finding might help to extend the general observation in social psychology that men tend to seek social status (Hewstone, Stroebe & Jonas, 2012) in that it provides an explanation at an ultimate level. This means that men might seek social status today because it would most likely have boosted inclusive fitness during our evolutionary past.

While arguments remain as to the importance of evolved predispositions in guiding our mate choice preferences, it is certainly the case that those ancestors who made good choices would have left more surviving offspring than those who made bad choices. We may today see around us flexible and culturally influenced marriage and courtship systems, but such flexibility is unlikely to be infinite or to lead us too far away from the predilections of our ancient ancestors.

So are men from Mars and women from Venus?

Unfortunately for those who wish to deny that there are evolved sex differences in human mate choice, the data really are quite overwhelming that they exist. Such people may gain reassurance from the fact that some of them are getting smaller as societies become more equal but there is no evidence of their vanishing completely. What is more, some of these findings are among the biggest in psychology. For example d is a statistic developed by Jacob Cohen to measure how big a difference between two means is. If you understand such things, a d of 1 means that the difference between two means is 1 standard deviation (which is generally considered to be a very large difference). If you don't understand, or care, about such things, then according to Cohen $d = .2$ (or less) is weak, $d = .5$ is moderate and $d = .8$ (or greater) is large. Most psychological research shows effect sizes of $d = .3$ (Hyde, 2005), whereas, for example, the sex difference in interest in casual sex is around .7 (Lippa, 2009) – so some of these sex differences are among the largest in psychology.

But what does "large" mean in this context? Does it suggest that men and women really are from different planets? This is a question that Steve Stewart-Williams and Andrew Thomas attempted to address by comparing sex difference size in humans with those in non-humans. Obviously, it is not possible to do this directly as non-humans cannot answer the questionnaires that are used in much of the research in humans. Instead, they measured this indirectly. Gibbons, for example, are generally considered to be monomorphic, meaning that males and females of these species are very close in size and behavior. Males are still bigger than females, just as for humans, but this difference is barely noticeable to the untrained eye. Yet, the effect size for the male–female size difference in size is .8 (Geissmann, 1993), about the same size as the "large" human sex difference for casual sex of .7. So even though .7 may be a large effect for psychological research, it is not, they argue, a large difference when we look across species, because it is approximately the same size as the sex difference in size for species that are generally considered to exhibit extremely small sex differences (Stewart-Williams & Thomas, 2013a; see Figure 4.6).

So, although sex differences are real, robust and culturally universal, it is possible to overstate the magnitude of this difference. Human males and females are much more alike psychologically than their closest relatives – gorillas, bonobos and chimpanzees. Certainly, not as distant as Mars and Venus. We will return to the debate about human sex differences in the final chapter.

Figure 4.6 Female and male white-cheeked gibbons. Note that although the female is lighter than the male, they are of very similar size.

Source: Image copyright © Timothy Craig Lubcke/Shutterstock.com.

Summary

- Individuals that are better able to make good mate choices are more likely to pass on their genes. Hence, selection pressures are predicted to enhance an animal's abilities to be selective on grounds of quality when it comes to choosing a mate. Isogamy is the state where two gametes (sex cells) are of the same size, whereas anisogamy is the state where the gametes are of different size. The two sexes evolved via anisogamy to take on different forms.

- Arising out of this differences in gametes males and females generally differ in their level of parental investment in offspring, with the latter investing more time and effort into the production of offspring than the former. There are four main types of mating pattern. In the case of monogamy, one male pairs off with one female, whereas in polygyny a male normally has access to a number of females. The reverse case – polyandry – is where one female has access to a number of males. Finally, promiscuity involves no long term commitment to a partner or partners by either males or females.

- Some experts have suggested that human males and females differ greatly in terms of their psychological make-up. Other experts take issue with this scenario and consider that such differences have been exaggerated. Evolutionary psychologists have made predictions based on evolutionary theory as to why males and females are likely to differ in their mate choice predilections.

- There are at least four reasons why we can predict that there may be differences between men and women in reproductive behavior. First, a man can, in theory, create many babies in the time that a woman creates one. Second, while women are fertile for a limited number of years, men are fertile until old age. Third, women invest more time and effort than men in the production of offspring. Fourth, only men can be cuckolded (bringing up another man's child).

- A number of predictions arise out of these asymmetries between the sexes. First, the reproductive value of a female will be more closely related to indications of youth and fertility then for men, which, in turn, means that men will find such indications attractive. Second, women will be more choosy then men and will seek indications of wealth, resources and social status. Third, men will seek younger female partners and women will seek older male partners. Fourth, males will develop strategies to avoid cuckoldry. Arising out of this the market values of men and women should be based on different criteria.

- Evidence from at least four types of study offer broad support for these predictions. A large-scale cross-cultural questionnaire study David Buss has uncovered evidence that males do prefer younger female partners, that women prefer men with resources and that males prefer attractiveness cues that correlate with youth and fertility. Likewise, through analysis of lonely hearts advertisements, Robin Dunbar has found the same pattern of age difference between males and females in requested romantic partners. Devendra Singh has uncovered evidence that men find a low waist-to-hip ratio (around .7) attractive in women. Singh considers this ratio to be an honest signal that males can use to assess fertility in females. David Perrett has found that females favor robust masculine features in a male during the fertile period of their estrus cycle, whereas they shift their preference to less robust men at other stages. Perrett suggests women may have evolved the possibility of a mixed-mating strategy by having extrapair relationships with more robust men and entering into long-term relationships with less robust men that may be more likely to make for good fathers (the mixing "dads and cads" strategy).

- The notion that human mating preferences can be related to evolved tendencies is not without critics. Biological philosopher David Buller, in particular, has offered alternative explanations for Buss's findings, suggesting the age gap between men and women in romantic relationships is a product of the fact that human females reach sexual maturity earlier in life than males. Evidence for this alternative hypothesis is not currently convincing.

FURTHER READING

Buss, D. M. (2004) *The evolution of desire* (2nd ed.). New York: Basic Books.
Milan, E. L. (2010) *Looking for a few good males: Female choice in evolutionary biology*. Baltimore: Johns Hopkins University Press.

Living with others
Evolution and social behavior

5

What this chapter will teach you

- Why humans evolved to be such social creatures.
- How William Hamilton's theory of kin selection has been applied to help explain self-sacrificing "altruistic" behavior in humans.
- How Robert Trivers's concept of reciprocation has helped us to understand cases of apparent altruism between non-relatives.
- How Trivers's theory of parental investment and of parent–offspring conflict can be used to predict when parents will provide care and when they are likely to reduce this level of care.
- How evolutionary psychology can help us to understand why we are susceptible to in-group–out-group biases.

Much of social behavior is prosocial behavior; that is, it involves responses that aid others. One of the problems for evolutionary psychologists is to explain, within a Darwinian framework, why humans provide aid for each other. As we saw in Chapter 1, since the 1960s and 1970s, evolutionists have helped us to understand at an ultimate level why we engage in apparently altruistic behavior. In this chapter we flesh out these ideas by examining the concepts introduced by William Hamilton and Robert Trivers. In particular, we consider the costs and benefits of living as a social species and examine studies that have tested Hamilton's theory of kin selection. We then move on to consider the work of Robert Trivers, including the concepts of **parental investment, parent–offspring conflict** and reciprocation in relation to human affairs. Finally, we consider why humans might be susceptible to act in potentially racist and xenophobic ways.

Evolutionary social psychology

Humans are an extremely socially integrated species – so much so that one of the worst punishments that can be meted out to anyone who has committed a crime is to place them in solitary confinement. Traditionally, social psychologists tend to ask quite specific questions about social behavior, such as, why do we form out-group prejudicial stereotypes (see later) or what is it about others that attract us to them (Hewstone et al., 2012)? They tend not, however, to ask why it is that we are such social creatures. Scientists trained in evolutionary theory (be they biological or social scientists) consider the costs and the benefits of **sociality**. Importantly, they also consider how and why prosocial behavior came to evolve given that **natural selection** supposedly promotes selfish behavior. Clearly, living socially can lead to greater safety from predators since more eyes improves vigilance and more bodies leads to improved defense. Also group living can improve thermoregulation in cold weather and social grooming can reduce parasite pressure (see Chapter 2). But sociality can also lead to greater competition for resources, and aggregations of animals can make prey species more conspicuous. These are the sort of fundamental questions that evolutionary social psychologists consider before then asking the same sort of questions (such as those above) that traditional social psychologists ask. From an evolutionary perspective, we would expect the benefits to outweigh the costs in terms of passing on copies of our genes if social behavior is to evolve (Alexander, 1974; Schaller, Simpson & Kenrick, 2006).

KEY TERM

Sociality The tendency of humans and animals to associate with others and form into groups.

The social behavior of our relatives

We tend to think of the **anthropoid apes,** such as gorillas, chimpanzees and bonobos (sometimes referred to as pygmy chimpanzees), as being very social species. It is certainly the

case that each of these great apes lives in a social group (varying in size from 12 to 100 individuals; Strier, 2011). In the case of chimps and bonobos, dominance hierarchies are formed where "everybody knows their place" in the group, whereas for gorillas social groups are formed around a single dominant silverback male and several mature females. But not all great apes live in social groups, as the predominantly solitary orangutan demonstrates. Orangs weigh in at 150 and 80 pounds for mature males and females, respectively. While females do at times form brief social groups with other females and their young, adult males and adolescents of both sexes live largely solitary lives. Orangs spend most of their time foraging for fruit high up in the treetops of the rainforests of Borneo and Sumatra (see Figure 5.1). This large size and arboreal lifestyle means that they are virtually free of predators. This is the main reason they do not live in social groups since they can feed more efficiently by spreading out and avoiding the cost of competition for food (van Schaik, 2004).

If our relatives the orangs can get along for the most part quite nicely without others of their species, then, we might ask, what led our ancestors

Figure 5.1 Unlike other highly socially integrated anthropoid apes, mature male orangutans give thumbs up to solitary living.

Source: Image copyright © Dr. Erica Cartmill. Reproduced with permission.

to become such a social species? Ancestral **hominins,** like gorillas and chimps, were of large body size but they spent far longer on the ground where they would have been vulnerable to large carnivorous predators. Fossil and artifact evidence suggest that, in fact, early hominins lived more out on the open savannah than do current living chimps and gorillas, so the safety in numbers argument may have been of particular importance to our ancient ancestors. As you may recall from Chapter 2, as the forests of East Africa receded, members of our ancestral species *Australopithecus,* finding themselves out on the open plains, would have become more vulnerable to big cat predators. The formation of increasingly larger groups no doubt aided protection from these predators – but our ancestors appear to have turned this enlarged group size to their advantage. As first *Homo habilis* and then *Homo erectus* evolved on the open plains of the African savannah, the increase in both brain and group size meant that in addition to increased protection, they also changed from scavengers to cooperative hunters (see Chapter 2; Dominguez-Rodrigo, 2002; Stringer, 2012). This highly cooperative behavior has been called *hominin mutualism* (Rossano, 2003). The complexity of **mutualism** in our ancestors is likely to have diverged from that of the great apes during hominin evolution in that coalitions grew larger and became more structured than other primates (Tooby & DeVore, 1987). This would have allowed members of *Homo erectus* to bring down prey much larger then themselves. Note that chimpanzee hunting parties, being smaller and less organized, are able only to bring down relatively small prey (Strier, 2011).

Cooperative hunting is only one of the driving forces that led to our species diverging from other primates. As we saw in Chapter 2, living in a larger, more complex social group also places a selection pressure to evolve a larger, more complex brain and hence a greater level of intelligence compared with the anthropoid apes. We will return to consider cooperation between non-relatives after first considering the importance of kin relationships.

Kin selection and social behavior

You may recall from Chapter 1 that the notion of kin selection was developed to explain altruistic or self-sacrificing behavior among relatives. We can formalize Hamilton's insight as an equation:

$$c < rb,$$

where **c** is the cost to the actor, **r** is the coefficient of relatedness and **b** the benefit to the recipient. This suggests that in all social species we can predict high levels of self-sacrificing behavior for close relatives (because ultimately this helps us to pass on copies of our genes). Moreover, Hamilton's equation also predicts that the closer the genetic relationship, the higher the level of sacrifice we are likely to observe. One way to think about kin selection is natural selection working at the level of the family. Under kin selection theory, individuals can pass on copies of their genes to future generations either directly (via offspring) or indirectly by aiding non-direct kin, such as nephews and nieces. Between them the direct and indirect paths to passing on copies of our genes add up to our **inclusive fitness**. Hence, to evolutionary psychologists, we humans are designed by natural selection to behave nepotistically in order to boost our inclusive fitness (or more precisely, due to natural selection, our psychological make-up is such that we are likely to engage in acts today that would have boosted our inclusive fitness in the ancestral past).

DISCUSS AND DEBATE – DECIMALS AND KIN SELECTION

Some social scientists have criticized the notion of kin selection on the basis that some cultures do not use fractions and decimals, suggesting they would find it difficult to determine who to provide aid for (and how much). How might an evolutionary psychologist respond to this criticism?

KEY CONCEPTS BOX 5.1

Kin selection in meerkats provides clues to human evolution

Due to their high level of cuteness, one of the biggest zoological attractions is surely the small social mammal that is the meerkat (also known as the *suricate*). When not amusing people at zoological gardens (or selling insurance on TV), meerkats are normally found in social groups of up to 30 individuals in southern Africa. In addition to their habit of standing up on their hind legs, another reason meerkats are such an attraction is the way that they stay close together and appear to look out for each other (which is one of the reasons they stand up). In the wild, meerkats not only look out for each other by scanning for the approach of predators and then providing alarm calls to aid others of their group, they also cooperate in three other ways – baby-sitting, social digging and pup feeding (a nurturing female will allow any pups in the group to suckle). Prior to Hamilton's kin selection theory, it was assumed that this self-sacrificing behavior was for the good of the group or species. Today it is known, however, that the members of each group are closely related, and the group is by and large an extended family of brothers and sister, aunts and uncles and cousins of various degree. Hence, by providing alarm calls, allowing others to suckle and digging holes for each other they are, in effect, helping to pass on copies of their genes and ultimately boosting their inclusive fitness.

Meerkats are not alone in this highly integrated behavior as, for example, lionesses, dwarf mongooses and the wild dogs of Africa, to name but a few, all engage in kin selected behavior beyond parent–offspring relationships (Clutton-Brock, 2009). Researchers have demonstrated quite clearly that in social animals where there is no cooperation in the rearing of offspring or in defending the group from predators, breeding success decreases as the group size increases. In sharp contrast where there is cooperation in the group on both of these behaviors, breeding success and survival both increase as the group size increases (Clutton-Brock et al., 1999). Hence, the sort of adaptations we find in meerkats and other highly cooperative social animals may help to provide clues to the evolution of group size and socially integrated behavior during human evolution. Over the last 40 years, it has become clear that kin selection theory really does solve the problem of most of the apparently altruistic behavior in social species.

Clearly, the notion that members of our species have evolved to be **nepotistic strategists** works well on paper. The question is does kin selection really operate in the real world for our species? When Hamilton originally proposed his theory of kin selection, it was intended to explain self-sacrificing behavior in animals such as social insects. Having established that kin selection is an important route to the passing on of genes during the 1970s and 1980s (see Key concepts box 5.1), evolutionists turned their attention to our own species. Over the last 30 years, evidence has accumulated that humans do demonstrate a strong tendency to favor kin over non-kin in the way that Hamilton predicted. The evidence of such behavior stems from a wide range of studies.

Nepotism begins at home

Imagine that you saw a runaway train heading toward a group of people, who, for some foolish reason, are standing on the line. These people are a mixture of friends and relatives of varying degree. Now imagine that you only had time to save just one of them – which one would you chose? When Eugene Burnstein and his colleagues posed such hypothetical life-threatening questions to a group of young adults, they found the results fitted in very well with Hamilton's prediction. People chose relatives over friends and the closer the relative the more likely they were to save them (Burnstein, Crandall & Kitayama, 1994). As we can see from Figure 5.2 not only were they more likely to aid relatives under life-and-death conditions but also the closer the relationship, the more likely they were to help out (remember the "coefficient of relatedness" (r) is a measure of the proportion of genes shared by common descent with 0.125 equaling a first cousin, 0.25 being equivalent to a nephew or niece, grandchild or grandparent and 0.50 being equivalent to a full sibling, offspring or parent). Participants also chose healthy relatives over unhealthy ones, young over old and premenopausal women over postmenopausal ones. All of these findings can be taken as suggestive that we may favor those relatives most likely to reproduce in the future – which would certainly be conducive with Hamilton's kin selection theory. Interestingly, when considering everyday favors rather than a matter of life and death, participants did not distinguish between relatives and non-kin to anything like this extreme degree.

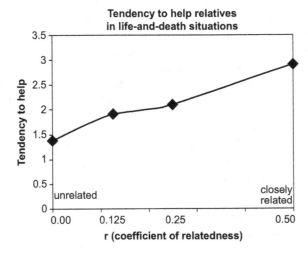

Figure 5.2 Self-report of how likely people are to aid kin and non-kin in hypothetical life-and-death situations.

Source: Adapted from Burnstein et al. (1994).

DISCUSS AND DEBATE – IF NEPOTISM IS NATURAL, SHOULD WE CRITICIZE IT?

Evolutionists suggest that we should expect to see nepotism in our species. Therefore, since it is "natural," should we accept it rather than consider it immoral and distasteful? Present arguments supporting or taking issue with this statement.

When social animals aid their kin should we label this nepotism, or should this term be reserved only for humans, with their awareness of moral codes?

Another approach to testing Hamilton's kin selection theory has been to consider historical archival data on behavior towards kin. One study that took this approach considered the role of kin altruism in Viking populations in Scotland and Iceland. Evolutionary psychologist Robin Dunbar (of "Dunbar's number" fame – see Chapter 2) and his coworkers Amanda Clark and Nicola Hurst found that, between the 10th and 20th century, alliances that were formed between kin were far more stable that those formed between non-kin (or in some cases, quite distantly related kin; Dunbar, Clark, & Hurst, 1995). Interestingly, kinship reduced the likelihood of committing murder to gain position and resources except when the potential benefits were very high. It might be suggested that when the benefits are very high, then kin selection breaks down – or, alternatively, that an individual might be able to produce a large number of surviving offspring by gaining large resources. If this is the case, then we might think of kin selection as simply operating more directly under such circumstances (that is, pass everything on to your offspring rather than spread your resources out to more distant kin). In a similar vein, many studies have demonstrated that in times of need, such as when orphaned or bereaved, kin are far more likely to help out than friends and the greater the level of aid required and the closer the genetic relationship the more likely the aid is to be forthcoming (Silk, 1980, 1990; Ivey, 2000; Bereczkei, 1998; Okasha, 2002; see also, Barrett, Dunbar & Lycett, 2002).

Do we have an evolved tendency to determine kinship?

If kin selection has been an important driving force in the evolution of *Homo sapiens,* then it will have been necessary for our ancestors to have evolved the ability to recognize kin and distinguish them from non-kin. From our present-day Western society perspective, the notion that we had to evolve tendencies to determine kinship might sounds a little odd. Surely, we know our kinship relationships with others because we are informed about this at an early age by people like our parents. We are, however, living in a very different environment to that of even our quite recent ancestors. The type of records we now keep and the ways of communicating instantly are recent inventions. During ancestral times, hominins would not have had such accurate (and immediate) access to this detailed kinship data (or, arguably, the linguistic/intellectual skills required to pass on such information). While individuals would not have moved very far away, given the lack of transport and of record keeping, there may well have been pressure on our ancestors to develop means of assessing the level of kinship. In fact, there are two evolutionary pressures that may have led to the development of an ability to assess our level of kinship with another to some degree of accuracy. First, if we are to behave altruistically to relatives, we would need to be able to assess degree of kinship in order to guide said behavior to the right people. Second, forming romantic relationships and breeding with close kin is likely to lead to genetic malformations in offspring (a situation known to geneticists as **inbreeding depression**). Clearly, those that were able to identify kin and hence channel resources more effectively would have been more likely to pass on copies of their genes than those lacking such abilities. Likewise, those most able to avoid incest-based relationships would also be more likely to pass on their genes to future generations, given offspring viability. Hence, natural selection would also have supported the ability to know who not form a romantic attachment to.

Stimulated by these two evolutionary pressures as to why it was necessary for our ancestors to have an understanding of kinship, Lieberman, Tooby and Cosmides (2007) have undertaken research that suggests we have evolved a **kin detection mechanism**. Prior to their work, it was generally regarded that humans, unlike members of other species, are quite poor at naturally identifying relatives (other than by being told about kin relationships). However, by testing 600 subjects on measures of which cues they use to determine and respond to kinship, Lieberman et al. have provided evidence that people compute a **kinship index**. Such cues include learning the features of those we associate with during development and using

KEY TERMS

Kin detection mechanism An evolved neural mechanism that is used to estimate a kinship index.

Kinship index A pairwise estimate of genetic relatedness between self and other.

Inbreeding depression The decline in vigor of offspring produced when two closely related individuals breed together.

these to match with others later in life. This "kinship index" can be thought of as an unconscious estimate of how closely related individuals are to us and involves quantitatively matching potential cues of relatedness.

Extending nepotism to ethnicity

Although the evidence that humans follow the predictions of Hamilton's kin selection theory appears to stand up well, we need to bear in mind that given our rich cultural heritage, other factors are likely to be involved in who we target when it comes to providing aid. In 2011, when studying decisions made by entrepreneurs, Chulguen Yang and his coworkers Stephen Colarelli, Kyunghee Han and Robert Page uncovered evidence that in addition to kin relationships, ethnicity can also play an important role here. Their study of Korean immigrants setting up businesses in United Sates demonstrated that close kin are supported when it comes to helping with business start-up funds. When, however, it comes to hiring workers, the ethnicity of customers trumped over kinship with once removed kin being treated no more favorably than others of Korean ethnicity. See Figure 5.3, and note that while those very closely related (r = 0.50) are given most aid those less closely related (r = 0.25 and r = 0.125), the r = 0.00 is largely accounted for by non-related Koreans.

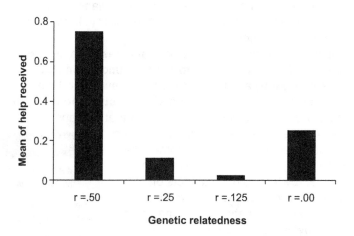

Figure 5.3 The mean amount of help provided for kin and non-kin in Koreans.
Source: Adapted from Yang et al. (2011).

Yang et al. suggest that in this case whereas a high level of assistance fits in with Hamilton's predictions, perceived nepotism may become extended in our species so that it appears to extend beyond kin to ethnicity. Perhaps unlike meerkats and lionesses, while humans have evolved, in part, to be nepotistic strategists, our cultural heritage is layered on top of this to extend "in-group boundaries" (see below).

Despite this, while Hamilton's hypothesis might have to be refined to a degree for our species, it is clear that kin selection has not been jettisoned, as kinship remains the most important predictor of associations between altruists and recipients (Maynard Smith, 1998; West, Griffin & Gardner, 2007).

Parental investment theory

Technically humans are a low-**fecundity**, long-lived species with a greatly extended period of development. This means that we have few children and each one takes an inordinate amount of time to develop to the point where it is able to leave the nest. Hence, the amount of support that we give to each individual child is enormous (even by comparison with meerkats). In 1972, Robert Trivers called the effort that parents provide for their offspring **parental investment** (see Chapter 4). In his words, parental investment is

> **KEY TERMS**
>
> **Parental investment** The amount of aid or resources provided to one offspring that might otherwise have been allocated to other offspring.
>
> **Parent-offspring conflict** Predicted conflict over resource allocation whereby the offspring demands more than the parent is prepared to provide at that stage in development.
>
> **Fecundity** The potential reproductive capacity of members of a species.

any investment by the parent in individual offspring that increases the offspring's chances of surviving at the cost of the parent's ability to invest in other offspring.

(Trivers, 1972, p. 55)

Note that Trivers was making it explicit that parents do not have an unlimited amount of resources and that deciding to invest in one child can reduce the amount they are able to invest in others (see Key concepts box 5.2).

Since its inception in 1972, researchers quickly came to realize that Trivers's concept of parental investment (PI) can be applied to help understand a lot of animal (and human) behavior. In particular, it helps to explain sex differences in behavior, since, when it comes to mammals, females nearly always invest more heavily than males. Think about it – female mammals have to feed the developing fetus when in the womb and then continue to do so through lactation following birth. Even in the most progressive of Western societies today, where the male is expected make a large contribution to child rearing, a man can never really even up the sheer cost of childbirth. And, of course, the form of aid that a father provides today in such societies is a relatively recent cultural invention. While human males, cross-culturally, do help out to a degree that surpasses 99% of mammalian species, and (judging by current living forager societies) have always been relatively high in PI, this always lags behind female PI.

With the development of what has become known as parental investment theory, Trivers realized that this asymmetry between the amount expended on offspring between the sexes explains why, as Darwin had suggested, females are choosy and males less so (see Chapter 4). Females are choosy because they bring so much to the relationship already and males are less so because high female PI can be seen as a resource that males will compete for. To put it in nutshell, because females invest so much in offspring they therefore need to be choosy over whom they recombine their genes with. For their part males attempt to impress females (and compete with other males) because females bring so much to the table. Trivers solved Darwin's problem with understanding why the driving force of sexual selection (female choosiness and male competitiveness) is this way around (see Chapter 1).

Why is parental investment important for understanding social behavior?

We have seen how asymmetrical PI between the sexes leads to differences in their behavior when it comes to selecting partners and that research in this area is so established that it has become known as parental investment theory. But what, you may be asking, has this to do with social behavior? The answer is that knowledge of PI helps us to both explain aspects of social behavior (within and outside of the family) and to make certain predictions about it. First, it allows us predict that, due to the lower costs and higher benefits, males will display more opportunism with regard to sexual partners than females. Second, it helps us to explain why families exist in the first place (e.g., they provide opportunities to invest in offspring). Third and finally, it allows us to predict when conflict is likely to happen within the family (see Key concepts box 5.2). Hence, any evolutionary-based understanding of social behavior requires knowledge of Trivers's parental investment theory (we will encounter PI once again in the final chapter).

KEY CONCEPTS BOX 5.2

Parent-offspring conflict

With his concept of parental investment in 1972, Trivers suggested that we can expect mothers (and, for some species, fathers) to invest heavily in their offspring. In 1974, however, he produced another ground-breaking concept that suggested there are limits to this level of sacrifice – and that when the offspring demands more than its fair share this will lead to **parent-offspring conflict** (Trivers, 1974). This sounds like common sense – but Trivers not only predicted conflict, he also used his own parental investment theory to predict when this conflict will arise. As the child develops the level of PI it wishes to receive will be different to the level the parent wishes to provide. This difference becomes increasingly large as the child becomes more independent. Trivers was able to pinpoint mathematically the time when conflict is most likely to peak. A prediction that has now been well borne out by a number of studies (Clutton-Brock, 1991, 2009). Hence, while parental investment theory can help to explain why families exist to boost their members' inclusive fitness, it can also be used to predict that family life will never be all sweetness and light.

Of the 650 MPs in the British House of Commons today, 20 are the offspring of former MPs. From a population of 60 million in the UK, this is many times higher than we would expect to find by chance. In societies across the world, we observe a degree of nepotism from the offspring of movie stars to the kin of politicians and of course the royal families of Europe and the Middle East. Although we don't approve of nepotism, we certainly expect to come across it in life. Hamilton's kin selection theory helps to explain this at an ultimate level of causation and Trivers's theory of parental investment adds further to our understanding of how and why relatives aid one another. And, yet, we are also surrounded by acts of kindness to non-kin – some of them above and beyond the call of duty. These require explanation.

One good deed deserves another – Dealing with non-kin

Greater love hath no man – Prosocial behavior toward non-kin

On April 15, 1904, steel tycoon and philanthropist Andrew Carnegie set up the Carnegie Hero Fund Commission to recognize extraordinary acts of heroism in civilian life. In the 110 years since its inception, the Commission has bestowed bravery awards to almost 10,000 recipients, all of whom had risked their lives to rescue another person. Recipients receive a medal with the Biblical words inscribed around the rim:

Greater love hath no man than this, that a man lay down his life for his friends.

Very rarely is the medal awarded for acts of heroism toward relatives. This fact might be taken as suggesting there are serious holes in Hamilton's kin selection theory. The opposite is true, however, as the commission defines acts of heroism as those where "the rescuer must have no full measure of responsibility for the safety of the victim." Such a definition of heroism would exclude many acts of sacrifice towards relatives, especially offspring. It might be suggested that the commission, in effect, recognizes that we expect relatives to engage in acts of self-sacrifice for each other with even life-threatening courage to kin not being defined as exceptional bravery.

While the Carnegie Hero Fund Commission discriminates against acts of self-sacrifice to family members, evolutionists still have to explain these acts and the other more mundane ones that we regularly engage in. If we evolved by Darwinian natural selection to become nepotistic strategists, why should we waste time and effort aiding non-kin? Once again, it was Robert Trivers who appears to have uncovered a plausible answer to this conundrum.

Figure 5.4 Carnegie Medal awarded for heroism.

Reciprocal back scratching

By the time he produced his groundbreaking work on parental invest-ment, Trivers was already famous in evolutionary circles for a theoretical paper he had published a little earlier. His paper, published in 1971, was simply called "The Evolution of Reciprocal Altruism," and in it spelled out how apparent altruism can evolve in unrelated individuals. Trivers con-sidered that in social species there may often be cases where one indi-vidual might take a small cost (or risk) in order to provide a large benefit for another individual. While this might not make sense in species where individuals rarely meet each other (and hence the best evolutionary strategy might be not to take this risk), in those that regularly encounter each other and are good at recognizing individuals, provided the favor is reciprocated, both parties can gain. The idea of a small cost (or risk) lead-ing to a big gain might sound odd – as you might think if I give you some-thing such as an apple and you give me an apple at a later date it simply evens out with no benefit to both parties. What we have to bear in mind, however, is that while we are living under conditions where an apple is a small treat, this was not the case either for most of our ancestral his-tory, nor is it currently the case for other species in the wild. In the natural environment one well-fed individual providing sustenance to another one that is starving can prove lifesaving. And if reciprocated at a later date, both parties would clearly benefit from the arrangement. (Note that the reciprocated aid does not have to be food, it might also be, for example, an alarm calling to warn of predators or social grooming to rid each other of parasites.) It is the cost/benefit asymmetry that forms the essence of Trivers's concept of **reciprocal altruism**. In setting out this new concept, Trivers laid out a series of conditions for its evolution in a species:

- The benefit of the altruistic act to the recipient should be higher than the cost to the actor.
- Animals must be able to recognise each other in order both to reciprocate and to detect non-reciprocators ("free riders").
- Animals should have a relatively long life span so that they may repeatedly encounter the same individuals.

Writing in 1985, Trivers suggested that reciprocity may have been very important in the evolution of *Homo sapiens* for three specific reasons. First, we clearly fulfill the three conditions set out above exquisitely; sec-ond, reciprocal behavior has been observed across all cultures and third, our species appears to have psychological adaptations to deal with recip-rocation. The third point is of particular importance since there is strong evidence that humans have evolved cheat detection modules (Cosmides & Tooby, 1992; Cosmides, Tooby, Fiddick, & Bryant, 2005). We will return to the notion of cheat detection when we consider the evolution of emotion and cognition in Chapter 7.

Since Trivers published his classic theoretical paper, a number of field researchers have made claims for the existence of reciprocal altruism in their social animal of choice (Clutton-Brock, 2009). These range from the gruesome exchange of blood between vampire bats (they are considered by some to take turns in regurgitating blood into each other's mouths when hungry) to reciprocal mating assistance in olive baboons (with subordinate males apparently taking it in turn to distract dominant males in order to gain access to females in season). Many of these have been disputed, however, and with the exception of some primate examples, reciprocal altruism appears to be relatively rare in the animal kingdom when compared with kin selected self-sacrifice (Clutton-Brock, 2009).

While the standing of reciprocal altruism has fallen in other species, its stock has simultaneously risen in importance for our own species. From anthropological studies that observe how forager people share food, to lab-based studies of games involving choices to cooperate or defect, time and again people across the world follow the credo that one good turn deserves another (Andreoni & Miller, 1993; Hill, 2002; Gächter & Falk, 2002; Storey & Workman, 2013).

The concept of reciprocal altruism has been refined since Trivers first referred to it in 1971. Today, it is often referred to simply as **reciprocation** to highlight that since the favor is returned it is not true altruism. Also, features such as **reputation** (where being observed in self-sacrificing acts by a third party can lead to benefits to the actor) and emotional responses such as sense of fairness and justice are now incorporated into it. We should also note that today some acts that might have previously been considered as reciprocation are redefined by some as mutualism (see Key terms earlier and Clutton-Brock, 2009). Despite these developments, it is clear that Trivers's original idea, albeit modified, really does provide an explanation of the social glue that binds non-kin individuals together to form societies.

Selfish or altruistic?

We would all, of course, like to think that we regularly behave in truly altruistic ways. And perhaps we do. But a lot of examples of apparently altruistic behavior may well have reaped benefits in our evolutionary past. It is entirely possible that reciprocation can even help to explain acts of heroism leading to the award of the Carnegie Medal, due to benefits provided by society to our surviving relatives. We should also bear in mind that while we may be able to rise above our naturally selected genes, most acts of kindness between non-relatives are small and when not reciprocated can result in serious resentment (Fehr & Gächter, 2000; Gächter & Falk, 2002). Such resentment towards free riders may then lead such non-reciprocators to either reassess their ways or be ostracized from social groups.

KEY CONCEPTS BOX 5.3

The Cinderella effect

Evolutionary psychologists Martin Daly and Margo Wilson have spent 40 years studying family conflict from an ultimate (evolutionary) perspective. One of their most shocking findings is that while the majority of stepparents do not harm their stepchildren, the chances of a child being physically harmed are greatly increased when a family contains a stepfather, with, for example, the rate of recorded infanticide being 120 times higher for stepfathers than for biological fathers (Daly & Wilson, 1998). In a book titled *The Cinderella Effect,* Daly and Wilson (1998) make use of kin selection and of parental investment theory to explain this difference in levels of abuse, suggesting that since stepchildren are not genetically related to their stepparents, they may not always release the kind of caring behavior that we normally anticipate from biological parents. In fact, as in the fairy story of Cinderella, such abusive stepparents might perceive the stepchildren as "cuckoos in the nest," taking resources from the own biological offspring. As can be imagined, the Cinderella effect has been disputed (e.g., Buller, 2005b), but Daly and Wilson have produced quite strong rebuttals to such criticisms (Daly & Wilson, 2005, 2007).

DISCUSS AND DEBATE – THE CINDERELLA EFFECT, THE LAW AND GENOCIDE

Can an argument be made to suggest that humans have evolved to provide more aid for biological children than stepchildren?

If so – can an argument be made that if a child suffers from neglect at the hands of a stepparent, courts should take this evolved tendency into account?

If humans have evolved to look after the interests of their kin, why do kin sometimes murder each other (known as genocide)? Make a list of predictions of the circumstances under which we might sometimes see genocide.

When compared with other acts of murder, should people who commit genocide be treated more or less harshly by the courts?

Out group hostility – The dark side of social evolution

As well as cooperating with kith and kin (kith, in case you ever wondered, means *friends* or *acquaintances*), like all other social species, humans have had an evolutionary history of competition for resources. Social psychologists have long known that the formation of a highly integrated group leads to the development of positive stereotypes about the characteristics of the group members. This helps to maintain group cohesion but unfortunately is at the expense of the development of negative stereotypes about members of other groups (Archer, 1996). Many experimental studies have confirmed this tendency that has become known as the **in-group–out-group bias**. This asymmetrical perception leads to favoritism for members of the in-group, but in contrast indifference or outright hostility to members of the perceived out-group. An example of this is Henri Tajfel's studies that involve dividing people up into two groups on completely random bases. They are then given the tasks of rewarding members of their own "in-group" and those of another "out-group" (both randomly created by Tajfel). Importantly, the participants apportioning the rewards do not get a share of the reward. Typically, the participants reward their own group more generously and, interestingly, describe their own group in more positive terms than the out-group to the point where the latter are seen as immoral and unpleasant (Tajfel, 1970, 1982).

KEY TERM

In-group–out-group bias A term used in social psychology and sociology to describe the general phenomenon whereby humans tend to perceive members of their own group positively when compared with members of other groups.

Many social psychologists today see this in-group–out-group bias as one of the seeds of intergroup conflict from school gangs to international warfare (Berkowitz, 2011; Krahe, 2001). Hence, gaining an understanding of this phenomenon transcends mere academic intrigue. Given that the in-group–out-group phenomenon is well established within social psychology, one question we might ask is how has the evolutionary approach added to this knowledge, and can it in any way help to reduce it? Interestingly, evolutionary psychology appears both to have added to our knowledge of the ultimate causes of racism and to have uncovered evidence that this is not inevitable. Kurzban, Tooby and Cosmides (2001) have suggested that humans may have evolved a tendency to gravitate towards in-group positive stereotyping and out-group negative stereotyping in order to boost coalitional affiliation. This in their view suggests that under some circumstances we may have a tendency to use ethnicity as a broad marker of in-group. This, in turn might lead to racism and xenophobia. Note that Kurzban et al.'s (2001) study fits in with the study by Yang et al., briefly outlined earlier, where ethnicity was seen as an extension of the concept of whom we should favor extending beyond family membership to ethnicity. Fortunately, while Kurzban et al. suggested that our coalitional tendencies can lead to racist attitudes they were also able to demonstrate how flexible people can be with regard to these attitudes. By changing boundaries of coalitional affiliation (i.e., Americans of African and European origin were presented as playing for the same team), they were able to over-turn racist categorization very rapidly. To Kurzban et al., labeling people according to race is not an evolved state but rather a "reversible byproduct of cognitive machinery that evolved to detect coalitional alliances." What they mean by this is that this is not a racist mechanism and such findings suggest to the authors that racism persists "only so long as it is actively maintained" (Kurzban et al., 2001, p. 15391; see also Cosmides, Tooby & Kurzban, 2003). What they mean by this is that this is not a "racist mechanism" since it is not designed to differ-entiate between races because when it evolved, your neighbors (against whom it was applied) would be of the same "race" as yourself. Rather, it is a mechanism that is applied nowadays to "race" (or ethnicity to be politically correct). Interestingly, most acts of racism/xenophobia in the world today are between near neighbors of the same original "race" or ethnicity (examples include Palestinians vs. Israelis and Catholics vs. Protestants).

Evolutionary psychology and social psychology today

Evolutionary psychology has provided social psychology with a new set of tools in order to derive new hypotheses and to explain old ones. Such tools include kin selection theory and parental investment theory (and the concept of inclusive fitness), which add the ultimate level of explanation to the traditional proximate level. Not all social psycholo-gists have warmed to the evolutionary approach (see Chapter 2) – but many have. According to three of these, Neuberg, Kenrick and Schaller (2010), having made inroads into our understanding of why we behave in prosocial and antisocial ways, evolutionary social psychology has now become a subdiscipline in its on right and one that is tackling an increasingly diverse range of topics. Such work is based on the notion that we have evolved adaptive psychological mechanisms to deal with recurring social challenges including, for example, social exchange, mate selection and disease avoidance. As Neuberg et al. suggest, however, because human social behavior is seen as adaptive does not mean that it is inflexible. Today, hypotheses derived from evolutionary psychology suggest we have predictably flexible responses to social cues (see, for example, the Kurzban et al. (2001) study briefly discussed above; see also Chapter 11).

In recent years there has been much debate concerning the notion of the extent to which humans have functionally distinct forms of psychological systems designed by natural selection to solve specific ancestral problems. Such mechanisms are known as *domain-specific mental modules*. We will examine this notion when we consider decision making in Chapter 6.

Summary

- Evolutionary psychologists have introduced the concept of kin selection and inclusive fitness to social psychology to help determine at an ultimate level why we act in prosocial and antisocial ways.
- Most species of anthropoid ape are highly socially integrated (the orangutan being much less so). When our early hominin ancestors left the forests for the more open savannah, group size may well have expanded rapidly for both protective and foraging purposes, leading to the highly cooperative state that has been called *hominin mutualism*.
- Hamilton's kin selection can be thought of as natural selection acting at the level of the family. Following kin selection theory, evolutionary psychologists suggest that humans have evolved to be nepotistic strategists. This means that in order to pass on more copies of our genes, we are more likely to aid kin than non-kin. A number of lines of research support this preposition including simulation scenarios where people have to make decisions about whom to save under various circumstances and archival records of acts towards kin and non-kin. Such studies suggest that cross-culturally people tend to support kin and that those most likely to go on to reproduce are given the most aid. Moreover, there is evidence that humans have a kin detection mechanism which aids us in determining kinship and that we are able to compute a kinship index which is an estimate of degree of kinship.

- Trivers developed the concept of parental investment – a measure of how much time and energy is allocated to one offspring that might otherwise have been allocated to other offspring. This has developed into parental investment theory since it leads to a number of predictions that can be tested. These include the notion that females typically invest more heavily in offspring than males, leading them to be more choosy when it comes to finding partners. Trivers also proposed the concept of parent-offspring conflict, which is the idea that the point when the parent wants to wean the offspring and invest in another comes at an earlier point than it does for the original offspring.

- Trivers also developed the concept of reciprocal altruism (also simply known as reciprocation). Reciprocal altruism consists of one animal providing aid to another in the "expectation" that it will later receive aid in return. It appears to be an important part of human social behavior, but there are debates concerning its role in other species.

- The Cinderella effect, which was proposed by Daly and Wilson, suggests that due to kin selection, stepparents are likely to lavish less attention on stepchildren than on their own biological offspring. While most stepparents do not neglect or abuse their stepchildren, the chances of these occurring is increased when the house hold contains a stepparent.

- The in-group–out-group bias refers to the tendency to perceive members of one's own group more positively than those belonging to other groups. It has been used by social psychologists and sociologists to explain hostility between groups. Evolutionary psychologists suggest that this propensity arose via natural selection to aid coalitional affiliation in the ancient past. While Cosmides et al. suggest that the in-group–out-group bias, if unchecked, can lead to racism and xenophobia, they have also uncovered evidence that such attitudes can be reversed when the boundaries of the group are altered and that racism only exists if it is actively maintained. This is an example of how evolutionary psychologists see our social behavior as flexibly adaptive having evolved to help us deal with various recurrent social problems from our ancestral past.

FURTHER READING

Hewstone, M., Stroebe, W. & Jonas, K. (2012) *Introduction to social psychology*. New York: John Wiley & Sons.

Neuberg, S. L., Kenrick, D. T. & Schaller, M. (2010) *Evolutionary social psychology*. In S. T. Fiske, D. Gilbert & G. Lindzey (Eds.), *Handbook of social psychology* (5th ed., pp. 761–96). New York: Wiley.

Schaller, M., Simpson, J. A. & Kenrick, D. T. (2006) *Evolution and social psychology (frontiers of social psychology)*. New York: Psychology Press.

Growing up
Evolution and development

6

What this chapter will teach you

- An overview of the nature–nurture debate explaining what it is and giving you different ways of thinking about it.
- Estimating the relative contribution of nature and nurture to individual differences using behavioral genetics.
- Early cognitive development: what babies can do.
- Life history theory as a way of understanding how genes might be able to tailor developmental trajectories to the environment.

Nature–nurture, development and evolution

The nature–nurture debate is one of longest standing and pervasive debates within psychology. In a nutshell, it asks the extent to which we are the product of our genetic heritage (nature) or our environment (nurture). Although the question is often phrased as if it is either nature *or* nurture that matters, most current theorists emphasize that both matter to a greater or lesser extent. Before we discuss the relative significance of each and its relevance of all of this to evolutionary psychology, it will help to consider in more detail what nature and nurture mean.

What do we mean by nature?

To the casual observer, the importance of nurture is obvious. Humans are born helpless. They cannot walk, talk and are dependent upon their parents (particularly the mother) for a considerable amount of time. This simple observation seemingly makes a mockery of anyone arguing that nature plays an important role in development. One might ask if language is *innate* – the word simply means inborn and is synonymous with nature – then why can't babies speak? This view represents a failure to understand what theorists usually mean when they discuss the innateness of some ability – a reasonable failure, as many theorists are unclear themselves as to what they mean when they say that something is innate.

Innateness is actually quite a clumsy concept, but it is one that we have been given by centuries of theorizing, so we have to deal with it. Here are some different conceptions of what innate might mean.

1   A baby is born with a particular ability or is capable, from birth of a certain behavior. We might call this *trait innateness.*
2   A baby is born with a particular set of predispositions that will eventually lead to the development of some behavior or attribute as the baby matures. We might call this *maturational innateness.*
3   A baby is born with a certain set of predispositions that, given certain environmental conditions, will lead to the development of behaviors or attributes later in life. We might call this **gene– environment interactionism.**
4   A baby is born with a set of potential developmental trajectories that, again, depending on the environment, can lead to following one life path or another. We might call this *environmentally contingent innateness.*

There are doubtless other things that people mean when they say such-and-such is innate, but these are the more common ones, and the ones that we discuss in this chapter.

Hopefully you can see that these different meanings can cause confusion. For example if we say that language is innate what we mean is something akin to version 3 – that we have a set of mental procedures that enable language to be learned as long as a child is exposed to language; if the baby is deprived of linguistic input entirely, they are unlikely to speak. However, it is quite understandable if someone thinks that our point is ridiculous because they think we are using *innate* in the first sense ("but newborns can't speak"). Likewise, puberty is innate in the second sense – you do not have to create a certain environment for a child to reach sexual maturity (although, as we shall see in this chapter, certain environmental factors might accelerate or delay the onset of puberty; it happens at some point). In case you think that the first definition of innate is a red herring, it is worth pointing out (and we shall see later) that newborns are capable of some behaviors. They are born being able to feed, and can react to a variety of stimuli, including pain, temperature, touch, gravity and – surprising as it may seem – faces and speech (although they don't understand the meaning of words).

Innateness is additionally clumsy because it creates an arbitrary dividing line at the point of birth. This is clearly a very important event for both the child and its parents, but it is worth remembering that development has been ongoing for around 9 months prior to birth, including, again surprisingly, learning about the world outside. For example, newborns attend more to their native language than an unfamiliar language with attention being measured by their sucking less on a pacifier capable of detecting the rate of sucking (Moon, Lagercrantz & Kuhl, 2013).

But how can babies have their "own" language when they can neither speak nor understand? It seems that they are responding to the sounds of the people around them when they are in the mother's uterus; this early exposure leads to the subsequent preference. So not only is finding that a baby *cannot* do something at birth not evidence that it is not innate (see points 1–4 above), but finding that a baby *can* do something at birth is not necessarily evidence that something is innate. It could be learned prior to birth, as our language example shows.

If we want a clearer definition of nature for the purposes of this book, then one which would encompass all of the above definitions would be "ultimately influenced by **genes**." Influenced rather than determined because, as we shall see later in this chapter, interactions between genes and the environment seem to be the norm rather than the exception. "Ultimately" because often the link between genes and behavior is indirect and sometimes tortuous in nature. Armed with a better way of thinking about nature, we now need to move on to nurture.

What do we mean by nurture?

Nurture comes from the same word as *nurse* (as in "care for"), so it is perhaps unsurprising that when people think of the influence of nurture they consider – as Freud did – the child's relationship with his or her parents. However, in more recent years, psychologists have come to consider nurture to mean the environment in the broadest sense (see Pinker, 1997). This is very broad indeed, because viewed this way nurture means "any influence that cannot be ultimately traced back to the genes." The reason for this is that given we accept that anything that influences us has to be either nature or nurture, once we have associated nature with the genes, nurture has to be every other influence. Here are some examples of sources of environmental influence.

- The way that the individual is treated by parents, peers and other people.
- Individual (i.e., non-social) experiences.
- Any physical traumas experienced in life (knocks on the head, oxygen starvation).
- Toxins.
- Diet and nutrition (or the lack of it).
- Disease.
- Experiences in the uterus, including placental problems, mother's diet, diseases, toxins, drugs and so on.

In Chapter 2 we discussed the various problems identifying the genes "for" particular behaviors, but this problem is relatively simple by comparison with identifying environmental influences: at least you know where genes are located and therefore where to look.

Nature–nurture: Explaining similarities and differences

A further important distinction that needs to be made with respect to the nature versus nurture debate is the extent to which we are attempting to explain differences between people or similarities. Before reading you might find it instructive to attempt the discuss and debate question.

DISCUSS AND DEBATE NATURE–NURTURE SIMILARITIES AND DIFFERENCES

Make a list of some questions that you find interesting as a psychology student. When you have done this, go through each and see whether it is concerned with differences or similarities (see next).

Psychology has a long tradition focusing on what makes people different from one another, and these seem to be of particular interest to psychology students. Some examples of this are as follows:

- Why are some people more intelligent than others?
- What makes some people shy?

- What is the cause of psychopathic behavior?
- Why do some people fail at school?

On the other hand, a considerable amount of psychology – especially evolutionary psychology – focusses on what makes some people similar. Here are some examples:

- Why are we disgusted by rotten meat, but less so by rotten vegetables?
- How do we learn language?
- Why do we forget things we want to remember, but remember things we would rather forget (e.g., in posttraumatic stress disorder).
- Why do we love our children?

Later we will look at some developmental accounts that focus on similarities, and later still we will consider some more complex interactions between genes and the environment (what we described above as environmentally contingent innateness). For now, we will discuss some attempts to separate nature from nurture in terms of individual differences (see Chapter 10 for more on this).

Twins and heritability

The problem with trying to separate out nature and nurture (well, one problem– as we shall see, there are many) is that there are so many "confounds." To remind you, the experimental method works by working out all of the possible variables that may have an effect on our target phenomenon, then systematically isolating each one so that we can determine which is important. For example, if I make terrible cakes and am interested in whether the reason is the flour I use or the butter I use, then I need to hold everything constant and look at the effects of each in isolation. If I were to change the flour *and* the butter and got great cakes, then I wouldn't know whether it was the flour or the butter that made the difference.

As it is with cakes, so it is with development. If I wanted to know why some people develop schizophrenia and some do not, and whether nature or nurture was most important here, then we are faced with a problem. If I gather together a sample of people with schizophrenia and a sample of people without, then all of those people will differ from each other in their genes (nature) and the environment. So, as for the cakes, I cannot decide which factor is important and which is not (or, more likely, which is more important).

You could imagine, in some nightmarish science fiction world, an experiment in which a group of babies were separated from their parents and brought up in minutely controlled environments: we hold the environment constant so any differences between people as to whether or not they develop schizophrenia is the result of genes. Or an equally terrifying world in which babies are cloned: we hold the genetics constant so any difference is environmental. Fortunately for us, we can do the second experiment, but before anyone considers ringing the papers about evil scientists, the cloning is not actually done by scientists; it is done by nature in the form of identical twins.

Identical or **monozygotic twins** are genuine clones in the sense that they are genetically identical. The product of a single fertilized egg that, for some reason, split into two: one individual is now two. Nonidentical, fraternal or **dizygotic twins** are the product of two eggs, each fertilized by a separate sperm. They therefore share 50% of their genes, just like regular siblings. The existence of monozygotic and dizygotic twins has allowed scientists in a field called **behavioral genetics** to do some important research on the **heritability** of certain traits.

KEY TERM

Monozygotic and **dizygotic twins** A **zygote** is the technical term for a fertilized egg. The prefix *mono-* means there was initially one fertilized egg and the prefix *di-* that there were two.

Estimating heritability

To remind you what heritability is: it is the extent to which variation across different individuals is accounted for by genetic differences. It is complex and fraught with difficulties, as we shall see, but the following is a crash course in the basic principles.

First, we need to think about **concordance**. Given a pair of twins, concordance is the extent to which the two twins are correlated on some trait. For example, intelligence. Take a sample of twins (either monozygotic or dizygotic), measure each twin using an IQ test and then look at the extent to which the scores are related across pairs of twins (which is expressed as a decimal between 0, meaning no relationship, and 1, a perfect relationship). If the results were entirely environmental, we would expect the concordance between monozygotic twins to be no larger than that between dizygotic twins. If the concordance between identical twins was 1 (perfect) and that between dizygotic twins much lower, then we could conclude that the behavior is probably genetic with little or no environmental input.

By combining the concordances between monozygotic and dizygotic twins we can get an **estimate of heritability** for the trait in question. As for concordance this is a number between 0 and 1, with 0 meaning that genes play no role in the variation across people (i.e., it is environmental) and 1 meaning it is purely genetic. This is sometimes converted to a percentage for public consumption.

A further source of data allows us to separate out environmental influence into two components: the **shared environment** and the **non-shared** or **unique** environment. As before, calculating this is complex but here, in essence, is how it can be done. The shared environment is the effect of growing up in the same family, so to estimate its effect we can compare identical twins reared in the same household with those who were adopted and reared in different households. If we find that the concordance for identical twins reared together is the same as that reared apart, then we can conclude (according to the logic used by behavioral geneticists) that the family is having no effect on the trait in question. If there is a difference, we can calculate the effects of the shared environment.

What about the non-shared environment? The total variation of a trait is described by the equation:

Total variation = effect of genes + effect of shared environment + effect of non-shared environment

We have already worked out the effect of genes, and the effect of the shared environment, and we know the total variation from the data we collected, so the effect of the non-shared environment is that which is left unaccounted for by the other two measures. There are other ways of doing this but this is, in essence, how the effects of nature and nurture are estimated.

What the data tell us about nature–nurture and individual differences

So, what do the results of behavioral genetics studies tell us? First is what is sometimes known as *Turkheimer's law* (Turkheimer, 2000): Everything is heritable. Turkheimer is a behavioral genetics researcher, and although his "law" might be a little overstated, it is surprising how many traits seem to have at least some degree of heritability. Most personality traits (extraversion, neuroticism, openness to experience – see Chapter 10) have heritabilities of around .5, meaning that 50% of the variation among people is due to genes. As an aside, it emphatically does not mean that for any individual 50% of their personality is due to genes and 50% is environmental, any more than if we find that height is 50% heritable (it is actually more than this) – that a 6-foot man inherits 3 feet from his genes and gets the remaining 3 feet from the environment – we are discussing variation among groups of people here. Intelligence is a little higher, with about .7 heritability (70%). Even behaviors apparently unrelated to genes are heritable. Voting behavior is heritable, suggesting that genes account for at least some of the variation in how one votes. But this can't be the case, since political parties and affiliations didn't exist in our evolutionary past. However, how one votes is probably partly due to personality: whether you are open to new experiences, for example, and since personality is heritable, it is not surprising that voting behavior is also. Rest assured, there is no gene for republicanism, per se.

What about the shared and non-shared environment? Rather surprisingly to many, much of the research shows that growing up in the same family has very little effect on trait variation. Or at least, it doesn't seem to make children any more similar to one another. For example, identical twins reared together are no more similar to one another than those reared apart, and unrelated adopted children are no more like each other than any two randomly selected individuals (see Plomin, 2011).

So, behavioral genetics research indicates that many of the differences among people have some kind of genetic basis. However, the simple partitioning of the origins of traits into this much nature and that much nurture is something of an oversimplification. Not one that is totally damning (although some may disagree), but one that is worth discussing.

First, there is the difficulty separating the effects of genes from the effects of the effects of genes. Before you close the book in frustration, read on – it really is simpler than you might think (see Key concepts box 6.1). Second, as we see next, heritability estimates are difficult to generalize from one population to another.

KEY CONCEPTS BOX 6.1

Gene environment interactions

First, consider that children who grow up with their biological parents are going to inherit some of their parents' traits and are also going to grow up in a household environment that reflects their parents' dispositions. Thus, intelligent parents will pass genes for intelligence on to their offspring but will also create an "intelligent" environment for their children, replete with books and ballet lessons. This double whammy is called a *passive gene-environment interaction*.

Second, consider that how you appear and behave? effects how people treat you. Friendly people tend to be liked, rude people disliked, quiet people ignored. If we take rudeness, then such people live in a world surrounded by people who act in a hostile way as a result of their rudeness. If we further imagine that rudeness is heritable (and it almost certainly is; see Turkheimer's law), then we can see that how people treat the person is an indirect effect of their genes. If the hostile reaction of people to our example's rudeness makes the person even ruder, then we can see that we have a vicious circle in which traits become more entrenched and stronger because the environment reinforces the traits that were initially caused by genes. This is a so-called reactive or evocative gene-environment interaction.

Finally, consider that people tend to seek out people who are like them and activities that satisfy them. So thrill-seekers hang out with other thrill-seekers and enjoy BASE jumping and other risky activities, again reinforcing the effects of the genes. This is called an *active gene-environment interaction.*

In these three examples we can see that the genes and the environment are acting together to reinforce a particular kind of behavior, but in behavioral genetic studies, because the behaviors are correlated with genes, they would all be put down as effects of the genes (or more strictly, the effects of the effects of the genes), even though the environment is playing a role.

Between groups, within groups and heritability

Finally, in our cautionary tale about behavioral genetics consider the example given by geneticist Richard Lewontin (1970). Lewontin asks us to imagine that we buy a normal packet of seeds from the grocery store. The seeds are normal seeds (not clones); all of them are genetically different from one another. We then plant the seeds in two planters as in Figure 6.1, but one planter receives normal concentration of nutrients and the other is deficient in nutrients. Imagine that we then ask, "Why are some plants taller than others, is it nature or is it nurture?" The answer depends upon which plants you are discussing. If we are asking why some of the plants within a particular planter are taller than others within the same planter, then the answer is that it is down to genes. The environment is the same for all plants within a planter, so the only possible source of variation is genetic. On the other hand, if we are to ask why the plants in one planter are on average taller than the plants in the other planter, then the answer is that it is environmental. Given that we randomly planted the seeds, there is no reason to suspect that there is any net genetic difference between the two groups, so it is almost certainly the difference between the two environments (nutrients) that is causing the difference.

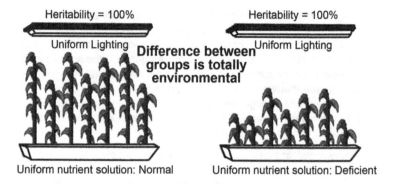

Figure 6.1 Example given by Richard Lewontin (heritability in plant growth – Ned Block). Source: Block, N. (1996). Reproduced with permission.

Enough about plants, what has this got to do with human development? Quite a lot, as a matter of fact. Turkheimer and colleagues (2003) reported on some research that looked at the heritability of educational achievement across several schools in the United States. The research found that achievement was moderately heritable. However, if one divided the children up into high and low socioeconomic status (SES), then the results indicated that for low-SES children, much of the variance in achievement was the result of the environment with genes having little effect, whereas for high-SES children, the situation was the other way round. It is not clear why this is the case. It could be that children from high-SES backgrounds all have such good environments (books, encouragement, extra tuition) that the only thing that makes one student better than the other is their genes; for the low-SES children, any extra benefits they get from the environment can make a lot of difference to their achievement.

Whatever the reason, the message becomes clear when we examine the findings through the lens that Lewontin helpfully provided for us: that the results of behavioral genetics research is, to a large extent, applicable to the group you have studied, and it can be erroneous to attempt to apply these results to other groups. This is important, particularly when the results of this kind of research are being used to inform public policy.

Summary: Is behavioral genetics research a waste of time?

Not at all. It has provided valuable insights, and compared to past research, which attempted to explain behavior in terms of the environment (usually) or the genes (rarely) while confounding both, it is a godsend. We genuinely think, therefore, that behavioral genetics research is a major contribution to understanding the cause of differences between people. However, we also believe that one needs to understand how such research is conducted and the various assumptions that are made, and the problems with these assumptions in order to fully appreciate it, and its shortcomings.

Cognitive development

Jean Piaget (1896–1980) is probably the most famous developmental psychologists of all. The problem that he focused on for most of his long life was how the intellect develops: how do apparently unintelligent babies become intelligent adults? In terms of the nature–nurture debate, Piaget was located toward the nurture side, believing that babies were born with basic knowledge and used this as a springboard to acquire more complex knowledge.

As an example of Piaget's reasoning, he believed that one of the most important tasks that the newborn faced was developing what he called the "object concept." This is the understanding that objects (including people) are enduring entities that exist independently of the infant's ability to perceive them. According to Piaget, this is not developed fully until the child reaches around 18 months of age, when they leave the **sensorimotor period** and move into the **preoperational period**. Piaget's evidence that this was the case was that children younger than 18 months would fail to search for a desirable object, such as a brightly colored ball, if it is subsequently obscured by an opaque screen. Piaget's interpretation was that to the infant, the object no longer existed.

Piaget's contention that infants know next to nothing about the physical world is a curious one from an evolutionary standpoint. Given that humans and their

Figure 6.2 Jean Piaget.

ancestors have had millions of years dealing with physical objects, and the behavior of physical objects has remained unchanged during that time (no changes in the laws of physics), it is odd that evolution did not give humans and their ancestors a head start by wiring such information into their brains.

The problem with babies

One of the problems with research into innateness – as mentioned above – is that human babies are so helpless. This is partly because all humans have to be born prematurely, when compared to other species. The bloated human head – a consequence of the large brain – and the constricted birth canal – a consequence of bipedalism – mean that babies need to be expelled from their mother's body before their heads get too big to fit. Even now, obstetric problems as a result of evolution's double whammy on the pregnant women are much more common in humans than in most other animals (barring a few pedigree dogs that also have too-large heads as a result of selective breeding).

KEY CONCEPTS BOX 6.2

A quick guide to Piaget's stages

Piaget's developmental theory is often referred to as a *stage theory*. What Piaget meant by developmental stages is often misunderstood and worth elaborating to prevent confusion. Piaget believed that children go through several revolutions in the way that they think about the world. Younger children in the *sensorimotor period* (0–18 months, approximately) believe that objects cease to exist when they are no longer in view. This is because they need to develop the object concept, in particular, object permanence, which is the understanding that objects are enduring entities. When infants develop object permanence they are said to be in the second stage, the *preoperational period* (18 months to 7 years, approximately). In this stage children have an odd understanding of matter that leads to them believing that the amount of "stuff" (say liquid or Play-Doh) can be changed simply by changing its shape (e.g., pouring liquid from a tall, narrow glass into a short, wide one changes the amount of water). This is known as a failure to conserve, and the experiments to show this are Piaget's famous conservation tasks.

Although often criticized for these studies, many children do show a failure to conserve, including one of the authors of this book who, as a child, used to crush up his potato chips in the bag because it "made more chips!"

Once a child is able to conserve, they are said to be in the *concrete operational period* (7–11 years, approximately), where they have a good understanding of the physical world but have difficulty with logic and abstraction. These are not fully mastered until the child enters the formal operational period at around 11 years.

There is a great deal of research on this (for and against) and many criticisms, not least Piaget's emphasis on solitary rather than social tasks and learning. If you are interested, turn to a developmental psychology textbook (see the Further Reading section).

The upshot of all of this is that newborns are so compromised in terms of their input and output responses that it is very difficult to determine what their brains are capable of. It is like trying to test how good a computer is, but with no keyboard and no screen, there's just no way of telling. This was the way things were until in the latter part of the last century, when some clever researchers found ways of capitalizing on things that babies, even newborn babies, can do, which is directing their attention and becoming bored.

Take object permanence. In the 1980s, Renee Baillargeon (1987) devised the cunning apparatus depicted in Figure 6.3. Essentially, babies of 3 or 4 months of age were seated in front of an apparatus in which a screen slowly rotated through 180 degrees away from the babies. Babies love moving things and were initially fascinated (denoted by visually fixating on the rotating screen). After a period they started to become bored, indicated by their starting to look at things other than the rotating screen, at which point a hand placed a block behind the screen at such a point that the screen would crash into the block and be stopped. At this point, one of two things could happen. As just described, the screen would rotate and be stopped by the block so that it was 112 degrees from the horizontal (the possible event). In another condition, the screen would rotate through its 180 degrees just as before *as if the block was just not there* (the impossible event). In both cases, the screen obscures the block before it should hit it and the second condition is achieved by having a trap door that the block invisibly falls into thus allowing the screen to rotate through its full 180 degrees.

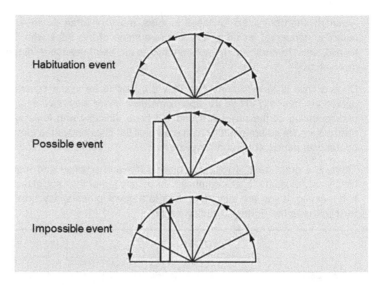

Figure 6.3 Baillargeon's apparatus for studying object permanence.
Source: Adapted from Baillargeon (1987).

So now, think about what would surprise you. And what would surprise the babies if Piaget were correct that they believed objects ceased to exist when they couldn't be seen. Have a go at answering these questions before you read on.

If Piaget were correct, then at the point at which the screen obscures the block, the baby should think the object has popped out of existence, and therefore should be surprised at why the screen is stopped at 112 degrees by nothing. In fact, just like you, it is surprised by the screen rotating through the full 180 degrees. It is as if it is thinking, "Where's that block gone?" Surprise is measured by looking time, more looking = more surprise. So Baillargeon's experiment indicates that at 3 months of age, babies have some expectation that objects continue to exist, even when they cannot be seen, in other words, they have object permanence 15 months earlier than Piaget thought.

Other research has shown that young babies have other competencies, including the knowledge that 1 + 1 = 2 and that 2 − 1 = 1 (Wynn, 1993), that objects cannot pass through one another or occupy the same physical space as another object ("no ghosts"), that an object passing from A to B has to pass through all of the points in between ("no teleportation") and that objects cannot influence one another without physical contact ("no telekinesis"). These last three principles discovered by Spelke, Breinlinger, Macomber & Jacobson (1992).

Note than none of this evidence conclusively shows that any of these physical principles are innate because they are all shown by babies of around 3 or 4 months, not newborns. What it does show is that such principles at least develop much more quickly than was previously thought, suggesting that they may be supported by innate principles (the second interpretation of innateness discussed above). Other research has shown that newborns have particular competences, including a preference for faces (Morton & Johnson, 1991), a preference for language-like sounds (Kuhl, Williams, Lacerda, Stevens & Lindlom, 1992) and the aforementioned preference for the rhythms and sounds of their mother tongue (Moon, Lagercrantz & Kuhl, 2013).

What is childhood for? Life history theory and development

This might seem like an unusual question, but it is a legitimate one if you are an evolutionist and is an example of evolutionary psychology questioning phenomena that other areas of psychology do not. Implicit in asking this question is that childhood has some kind of function: it would make little sense, for example, to ask what a pebble is for as it is not something that – in the natural world at least – has a function. First of all, let us extend the concept of childhood so that it encompasses non-human animals by defining it as a period when an animal is unable to reproduce sexually and, as we saw in Chapter 1, evolution is all about reproduction. The problem is why should evolution allow children the luxury of living for 12 to 14 years in the case of humans while being unable to reproduce? Or to put it in gene-centered language, if a mutation arose that led to humans maturing much earlier that would surely permit them to have more offspring, and therefore potentially outcompete genes that led to slower maturing bodies. Given that this has not happened, we should assume that there is perhaps a purpose for this long process of maturation (our close relatives, the great apes, reproduce at half the age of humans).

> **KEY TERM**
>
> **Comparative psychology** A branch of psychology that attempts to understand humans by comparing them to other organisms, usually mammals but other animals too.

Perhaps some help might be obtained if we think comparatively. A butterfly spends its childhood as a caterpillar. The caterpillar forgoes sex in order to concentrate on its primary activity, which is to eat (you might know some people like this). All its adaptations from its powerful jaws to its expandable body are designed to put on weight and avoid predation through camouflage, by resembling scary monsters by having fake "eyes" (see Figure 6.4), or being poisonous.

Figure 6.4 Caterpillar with fake eye spots to resemble a snake or other feared animals in order to deter predators.

Source: Photo by Michael Hodge, https://www.flickr.com/photos/mhodge/1216047199/ [CC BY 2.0 (https://creativecommons.org/licenses/by/2.0/], via Flickr.

This is all in preparation for the next stage because, as every child knows, the caterpillar will eventually become a pupa, in which all its body parts are broken down into their constituent parts – it essentially digests itself – and are reassembled to form the structure of a butterfly. Butterflies are designed for one thing: sex. And their adaptations reflect this. Eating becomes secondary to sex (you may know people like this too); instead, many sip sugary nectar, which provides energy for flight but little in the way of nutrition for bodily repair. They have wings for the same reason maple seeds have wings, to aid **dispersal** so that offspring are not competing with one another, or with parents, for food and other resources. They are often brightly colored to attract each other, and many species produce chemicals called **pheromones** that are also designed to attract members of the opposite sex.

> **KEY TERMS**
>
> **Dispersal** A key problem for many organisms is to prevent overcrowding among their offspring and therefore competition for food and other resources, which can lead to increased mortality. Therefore, many organisms have adaptations for ensuring that their offspring are spread far and wide. Examples being the wings of adult insects and the rotors of maple seeds.
>
> **Pheromones** Hormones that are used to affect the behavior of other organisms. Often used as a mechanism for sexual attraction.

Back to humans

But humans aren't butterflies, so what is the relevance of this to human development? Well, it gives us a way of answering the question: What is childhood for? Childhood is ultimately a period that prepares us for adulthood but at the same time it contains adaptations that are specifically designed for being a successful child rather than a successful adult (Harris, 2009). This might seem confused, but think of butterflies. The long-term success of the adaptations of a caterpillar are determined by its ability to become a good butterfly and have offspring – caterpillars cannot reproduce themselves, so this is the only way that good adaptations can be passed on to the next generation. But, equally, many of the adaptations – camouflage, eye markings, mouth parts – are specific to ensuring that the caterpillar survives. This view of childhood is known as **life history theory,** and we shall look at this approach to human development, next.

A life history theory of development

A great deal of research in developmental psychology has focused on the way that the environment affects subsequent development. Attachment (Bowlby, 1969; Ainsworth, 1967) is one of the most familiar examples. Attachment theory has been modified much over recent years, but in its earlier conceptualization it attempts to explain how relationships develop by tracing them back to the earliest relationship that a child has with its primary caregiver (usually, and originally, the mother). It was supposed that as a result of differing levels of material sensitivity, children could develop either a secure or various forms of insecure attachment. For example **insecure-avoidant** individuals find it difficult to get close to people and people find it hard to get close to them. At the opposite end of the continuum, insecure **anxious-resistant** (or **anxious-ambivalent**) individuals are anxious that they will lose the love of their significant other, often leading them to becoming angry with the person they love, or engaging (when they are adults) in partner-monitoring behaviors.

As the terms secure and insecure imply, these styles of attachment can have implications for future relationship success, with securely attached individuals tending to have more stable, longer-term relationships than either the insecure-avoidant individuals, who tend to have less intimate and more superficial relationships, and those with anxious-resistant relationships, who tend to have more fractious and volatile relationships. Insecurely attached individuals also seem to have more sexual partners, have more relationship breakdowns and tend to show an interest in sex earlier than their securely categorized contemporaries.

DISCUSS AND DEBATE – TEENAGE PREGNANCY

An issue that often receives media attention is the phenomenon of young teenage girls becoming pregnant. Discuss some of the reasons for why some girls become pregnant at an early age and some do not.

One of the many puzzles about attachment styles is just how many children and adults are classified as having one of the insecure attachment styles: around 25%. If, as the name implies, secure attachments are "normal," there seems to be lot of abnormal people out there. One possible explanation for this is that the traditional, stable, long-term, monogamous relationship is only one of many strategies that can be used in romantic relationships. Remember that in Chapter 1 we discussed that for humans there are broadly three areas that adaptations can be directed toward: survival, reproduction and ensuring that offspring survive to reproductive age. Focusing on offspring survival, we saw two strategies, actually better described as a continuum, known as **K-selected** and **r-selected** strategies. Organisms that adopt the r-selected strategy have many offspring but invest little in each one (think of fish, fungi or turtles), whereas organisms that adopt the K-selected strategy produce much fewer offspring but invest more in each (think of mammals such as chimpanzees, who engage in protracted childcare). One of the crucial factors determining which strategy is adopted is environmental riskiness. If there is a high probability of infant or parental mortality, then r-selected strategies tend to be adopted by a species; lower risk tends to lead – by evolution – to the K-selected strategies. The reason for this is simple economics: there is little point a parent evolving adaptations for costly childcare if there is a high probability that the child is going to die as a result of the risky environment. Likewise, it will do an offspring little good to be dependent upon a parent that is likely to die. Thus, harsh conditions can lead to little or no childcare and offspring who are independent from birth.

A more general way of thinking about r- and K-selection, and one that is easier to remember, is whether organisms are designed to maximize current or future reproductive success. Maximizing current reproductive success includes the r-selected strategy where the animal simply produces offspring with no plan to invest in them in the future. Maximizing future reproductive success includes the K-selected strategy, where individuals have children with a plan for long-term investment. Adopting a strategy that maximizes current reproductive success is a sensible response to a risky environment. There is little point investing in a future that, for you, might not exist.

Although we have described maximizing future or current reproductive success as two different strategies, it is better to think of them as a continuum, with some organisms being toward the "current" side of things and others toward the "future" side (Chisholm, 1999). For this reason it is known as the **C-F (or current-future) continuum**.

The previous discussion has concerned different strategies adapted by different species, but an interesting twist on this has been proposed by researchers such as Jay Belsky, Henry Harpending and James Chisholm, who argue animals – including humans – can shift along the C-F continuum, depending on environmental conditions (Chisholm, 1999).

Let us consider how this might work. We assume that from an early age an organism is sensitive to environmental risk; if it detects a high risk environment it adopts a strategy that maximizes current reproductive success, becoming sexually mature more rapidly and having as many offspring as it can and investing comparatively little in each. It behaves more like a typical r-selected organism. Alternatively, if the organism detects a low-risk environment, then it can develop more slowly, acquiring the skills (see Key concepts box 6.3) necessary for adulthood, and are not in so much of a rush to have offspring so tend to mature more slowly and have children later in life.

KEY CONCEPTS BOX 6.3

A life history account of play

Many young animals, especially mammals, engage in play, and for centuries there were debates as to what this apparently functionless activity was for. Why should it be for anything? Well, from an evolutionary perspective, organisms that spend time and energy engaged in pointless activities should suffer (in terms of **natural selection**) compared to those that use their time and energy judiciously. The consensus now is that play, far from being pointless, is a vital part of development. Juveniles are, in fact, practicing the skills that will become useful in adulthood but in a safe environment: kittens play at hunting, young male chimpanzees play at fighting and so on. According to life history theory, then, play is an activity designed to maximize future reproductive success because its benefits will only be realized later on down the line.

But, the extent to which an animal engages in play is conditional upon environmental conditions. When riskiness is high, animals play less (Fagen, 1977), presumably because they are spending time engaged in activities that maximize current reproductive success. Not only can risky environments lead to lower levels of play, they can change the kind of play that the animal engages in. Bateson, Mendl and Feaver (1990) found that female cats that were on a restricted diet weaned their offspring more quickly than those on normal diets. This is to be expected as the cats want to discharge their burden of childcare as quickly as possible under poor conditions – they are maximizing current reproductive fitness. Once weaned their offspring play, in fact they play more than the offspring of cats on normal diets, but the play is different spending more time practicing hunting – important if they are to achieve independence – at the expense of social play which is associated with more long-term goals.

Figure 6.5 Brown bear cubs, like humans, engage in play.
Source: Image copyright © Bildagentur Zoonar GmbH/Shutterstock.com.

So, what we have here is a theory which suggests that natural selection has endowed individuals with the ability to tailor their reproductive strategy in accordance with environmental riskiness in order to make the best of their situation. Is there any evidence for this in humans? In fact there is, although this is a relatively new area of research, so at present it is limited.

Research shows that signals of riskiness can lead to early sexual maturity, in girls at least. Belsky, Houts and Fearon (2010) found that 65% of girls categorized as insecurely attached at 15 months of age experienced menarche (onset of puberty) at less than 10.5 years of age compared with 54% of those categorized as secure at the same age. Belsky et al., (2010) found that girls whose mothers treated them harshly when they were 15 months of age became sexually mature earlier as before but also showed greater sexual risk taking when they were adolescents. Measurement of maternal harshness was achieved by the responses that their mothers gave to a questionnaire. "Harsh" mothers, by the way, are those who spank their children for doing something wrong, believe that children should respect authority and be quiet when adults are around, believe that praise spoils the child and do not give many hugs.

Risky environments are also associated with early pregnancy. A study of 4,553 British women in the United Kingdom by Nettle, Coall & Dickens (2011) found that early pregnancy was accounted for by environmental risk factors such as low birth weight, short duration of breastfeeding, separation from mother in childhood, frequent family residential moves and lack of paternal involvement. Socioeconomic status and the age of the participant's mother when they were born were controlled for in this study.

You might be thinking, What about boys? Are they unaffected by the environment? The answer is almost certainly no, but so far little research on boys has been forthcoming. One reason is that menarche – the age at which a girl has her first period – is such an easy and unambiguous way of measuring the onset of puberty. By contrast, there is no such neat marker for boys. Even an apparently simple indicator, such as the breaking of the pubertal male voice, is not so clear-cut as it can take a period of time to complete. Males also don't get pregnant (it is hardly worth saying), so there is also no objective indicator or early sexual experience. We could, of course, ask them, but boys are notorious for making up early sexual encounters and overestimating the number of sexual partners that they have had (Brown & Sinclair, 1999).

It is important to stress that the life history account of development argues that we should rethink our understanding of environment effects on development. Previously (e.g., in attachment theory), the environment was treated as a force that, together with but independent from genes, shaped the way that humans turn out. This view suggests that one of the effects of genes is to make organisms that are sensitive to different kinds of environment and can put into an effect a particular plan of action for dealing with that environment. This should also give the lie to those who think that evolutionary psychology in any way ignores or downplays the environment. Evolutionary psychologists should take the environment very seriously indeed because, as the evidence above is starting to show, our genes certainly do.

Summary

- Research in development has been dominated by the nature–nurture debate, with researchers typically emphasizing the importance of one over the other. Recent research shows that both genes and the environment play a significant role in human development.
- Although seemingly simple, there are many different ways that genes can exert and influence on the developing child. Some of these are direct, for example, genes for temperament, eye color and so on, whereas other indirect genes that affect someone's skin color may affect how people treat the child, which can have implications for development.
- Studies looking at individual differences in development using behavioral genetics show that nature and nurture both have a role to play in many behaviors. But it is important to understand the limitations of such research.
- Much recent research looking at similarities in development show that babies are capable of much more than we might imagine, providing support for the notion that the development of these abilities is innately guided.
- The life-history approach to development sees childhood as preparing the child for adulthood. Children can follow one or other developmental trajectories, depending on early experiences.
- Recent research suggests that environmental riskiness can have effects on the reproductive development of humans and other animals.

FURTHER READING

Block, N. (1996) How heritability misleads about race. *The Boston Review*, 20(6), 30–35. (Available from http://www.nyu.edu/gsas/dept/philo/faculty/block/papers/Heritability.html). A brilliant five-page article that cuts through the mystery of behavioral genetics research and, in particular, controversial work on race and IQ. If you read this and enjoy, search out his longer article published in the journal *Cognition*.

Chisholm, J. (1999) *Death, hope and sex: Steps to an evolutionary ecology of mind*. Cambridge: Cambridge University Press. An overview of life history theory as it applies to human development.

Gopnik, A., Meltzoff, A. N. & Kuhl, P. K. (2001) *How babies think: The science of childhood*. London: Phoenix. An excellent introduction to some of the recent work on cognitive development.

Decisions, decisions

The evolution of cognition and emotion

7

What this chapter will teach you

- Although cognition and emotion are different entities, they work together to enable us to make decisions relevant to inclusive fitness.

- Perceptual, reasoning and memory systems are optimized for making good decisions rapidly, such as enabling navigation or dealing with free riders.

- Some aspects of cognition (e.g., memory) may reflect the kinds of decisions that our hunter–gatherer ancestors needed to make rather than the decisions of 21st-century humans (e.g., protecting ourselves against predators).

- Both positive and negative emotional states arose to aid survival and reproduction.

Cognition and emotion: Two things or one?

One of the more familiar dichotomies in psychology – and indeed in everyday life – is that between thought and the emotions. The former is seen as being rational and cool-headed and the latter irrational, hot-headed and passionate (in days gone by, the emotions were referred to as "the passions"). Although this distinction is very old indeed, the separation of the rational mind and the emotional body is usually attributed to the French philosopher René Descartes. We will discuss some of Descartes's other ideas shortly, but this separation of the intellect and the passions needs to be discussed in order for us to explain why we have paired such unusual bedfellows in this chapter.

What is cognition?

When we think about cognition we usually treat it as synonymous with thought, and indeed the term *cognition* comes from the Latin word *cogito*, which means "to think." But when cognitive psychologists discuss cognition they mean something much broader than the kind of conscious, deliberate activity that we often associate with thought. For example, in addition to conscious thought, cognitive psychologists also study unconscious, automatic process, such as those that underlie visual processes. So, a definition of cognition could be "any mental activity in which information is stored, retrieved or transformed." Hence, the storage and retrieval of information that constitutes memory involves cognitive process, as does the way in which visual information is processed to render light waves into the perception of three-dimensional objects.

So, cognition is much more than conscious thought, and its principles can be applied to any kind of information processing system, from explaining the behavior of a computer to the behavior of a simple (presumably non-conscious), entity, such as an amoeba.

What is emotion?

We consider emotion in more detail later, but generally an emotion can be thought of as a particular mode into which an organism switches in response to particular internal or external situations in order to promote an appropriate behavioral response. For example, a threat to an organism's life might lead to it switching into a mode that leads it to rapidly remove itself from the threat, which we might call "fear"; the potential for food or other desired resources might lead it to switch into exploration mode, which we might call "curiosity." In many animals, these mode switches are accompanied by conspicuous physiological changes, such as sweating, raised heart rate, dilated pupils and in some animals, signals such as bodily or facial expressions.

Why we need cognition, why we need emotion

We might ask, and many have, why we need to have emotion at all. Why don't we, as humans, just have cognition? If we are faced with a threat, why don't we simply make a cool, rational decision to withdraw from the situation and act accordingly? The answer is that emotional responses have profound effects on the efficiency of the behavior. Again, consider fear. Imagine a hungry animal that has just settled down to eat. Suddenly, a predator appears that might endanger its life. A purely cognitive animal might weigh up the pros and cons of running away with the possibility that it will starve through lack of food, against staying and eating the food with the possibility that it might get be killed. While it is reflecting on this conundrum, the predator attacks and the decision has been made for it.

Armed with a fear response, the animal is rapidly put into a mode that increases vigilance, a physiological effect of the fear response is for it to lose all interest in ongoing activity and divert all of its attention (a cognitive effect) to the predator. Other effects are an increase in blood-sugar levels for energy and the diversion of blood from non-essential processes, such as digestion, to the muscles that enable fight or flight. It is now like the proverbial coiled spring. The predator strikes, the animal is already gone.

Later, when information processed from the sense organs inform it that there are no threats around, the fear response subsides, and the animal settles down. Receptors in the animal's gut detect that food is needed, and so it is switched back into exploratory mode in order to find food, attention is now diverted to attending to this activity and the animal starts to forage.

So, cognition and emotion are intricately entwined. Cognitive processes inform the animal of what is going on in the outside world – Are there threats? Is there food? – and the inside world – Am I hungry? Am I thirsty? – where necessary; these lead to emotional responses – Should I flee, fight or feed? – which prepare the body for appropriate action and also prepare cognition by diverting attention toward certain stimuli and away from others.

In this chapter we flesh out how evolutionists have contributed to the understanding of cognition and emotion by considering them in terms of their function on the organism. In other words, how do cognition and emotion enhance survival and reproduction?

Cognition

Prior to the arrival of cognitive psychology, the dominant psychological approach in the United States and the UK was **behaviorism.** Its chief proponent at the time, Burrhus Frederick (B. F.) Skinner, believed that

one should focus only on the stimulus and behavior and should not consider the processes that intervened. This is sometimes known as the **black box** approach because it treats the mind or brain as containing mysterious processes that cannot be understood. Skinner's reason for adopting this approach was that to do otherwise was unscientific. Because science, as he saw it, has to focus on observable phenomena and we could not study the mind or brain directly, to attempt to do so would be unscientific.

Figure 7.1 B. F. Skinner.
Source: Silly Rabbit [CC BY 3.0 (http://creativecommons.org/licenses/by-sa/3.0)], via Wikimedia Commons.

DISCUSS AND DEBATE

To what extent do you think that the mind is like a computer? What are the similarities and what are the differences?

KEY TERM

Computer models A computer model is a computer program that is designed to mimic the behavior of a human on some specific task.

The early cognitive psychologists such as Alan Newell, Herbert A. Simon, George A. Miller and Donald Broadbent thought differently. They drew inspiration from the digital computer, which was becoming increasingly commonplace in the 1950s. Like the human mind, a computer took in information, did something with that information and performed some action on the basis of that information. Of course, we do know what happens inside the computer because it was built and programmed by someone, so if we can build computers that behave is a similar way to humans, then it might tell us what is going on in the human mind. In some of these cases, **computer models** were not actually made; instead, a set of potential process was specified as a flow chart: the famous boxes and arrows diagrams (e.g., Broadbent, 1958, see Figure 7.2).

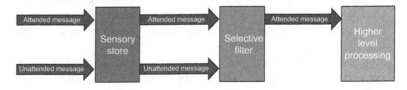

Figure 7.2 Broadbent's filter theory of attention.
Source: By Kyle.Farr (Own work) [CC BY-SA 3.0(http://creativecommons.org/licenses/by-sa/3.0)], via Wikimedia Commons.

Others, such as Newell and Simon (Newell, Shaw & Simon, 1959) built actual programs that were designed to mimic human problem solving.

Although cognitive psychology has progressed a great deal since the 1950s, its key tenets remain the same: to understand the mind and brain by considering it as an information processor. Perhaps the most important advance has been the arrival of **neuroimaging techniques** that allow researchers a glimpse inside the black box.

KEY CONCEPTS BOX 7.1

Neuroimaging

A set of techniques sometime called *brain scanners* that enable researchers to see what is happening in the brain. Examples include PET (positron emission tomography), CAT (computer axial tomography), MRI (magnetic resonance imaging), fMRI (functional magnetic resonance imaging) and MEG (magnetoencephalography). The latter two are particularly important because they enable the brain's activity to be captured over time as it actually changes in response to a given stimulus. The others present a static "snap shot" of the brain as it processes stimuli. There are differences in the way that these methods work – some subtle, some more marked. CAT (or CT) scans work by firing a very large number of x-rays from many different angles through the brain, which are then assembled to form three-dimensional images.

PET scans require the participant to ingest a solution containing radioactive glucose that emits particles known as positrons. When performing a cognitive task, the regions of the brain responsible use glucose for energy, and the positrons emitted are detected by sensors, enabling researchers to determine which parts of the brain are active.

MRI, fMRI and MEG rely on the fact that oxygenated and deoxygenated hemoglobin have different magnetic properties. Using sensitive detectors, these differences can be detected, enabling the degree of blood flow in the brain to be measured and thus determine which parts of the brain are more active on a particular task.

A final method not strictly speaking neuroimaging but nonetheless important is transcranial magnetic stimulation (TMS), which uses powerful magnets to stimulate certain areas of the brain with the researchers observing the effects on behavior. This technique can also be used to shut down some brain regions to create a temporary artificial lesion, similar to that received following a stroke.

Evolution and cognition

Although cognitive psychology is principally focused on *how* we solve something rather than ultimate questions such as *why* we solve something in a particular way, considering cognition from an evolutionary perspective certainly has the possibility to clear up some mysteries. Some examples of this can be seen in the study of visual illusions.

Evolution and the visual system

Look at the picture in Figure 7.3 (from Adelson, 2000).

You probably won't believe this, but the squares labeled A and B are exactly the same shade of gray. Proof is given in Figure 7.4.

The two vertical lines are exactly the same shade throughout their length (cover up the rest of the picture if you don't believe it); yet, each line blends in perfectly with the two squares.

This seems to show that our visual system is "buggy." If that is the case, then how do we get by with a visual system that is so poorly designed that it makes us see differences where there are, in fact, none?

Edward H. Adelson

Figure 7.3 Do the two squares, A and B, look similar or different?
Source: Adelson (2000). Reproduced with permission.

Edward H. Adelson

Figure 7.4 Proof that they really are the same shade of gray.
Source: Adelson (2000). Reproduced with permission.

DISCUSS AND DEBATE – WHAT IS THE MIND FOR?

Think about your bodily organs. They all have clear functions: the heart is a pump, the kidneys filter the blood and the stomach digests food. Try and think of the ultimate function of the brain in terms of what it does for survival and reproduction.

According to the British neuroscientist and psychologist David Marr (1982), examples such as those above show that the visual system is, in fact, *well* designed. To see why, look at Figure 7.3 again. What you see is a checkerboard "floor" of squares of two shades with a shadow falling across it. If, however, your visual system faithfully reproduced the light intensities hitting your retinas, it would look like a checkerboard pattern with squares all of kinds of different shades (look at the shades at the edges – they are different to the surfaces, for example) and if the shadow were to move it would appear that the squares were changing as it passed over them. In short, the world would change dramatically from moment to moment as lighting conditions changed. So, this bug, as software engineers are apt to say, is not really a bug of the visual system but a feature. The visual system is able to create a stable word in the face of large and rapid changes in luminosity.

Why is the visual system designed this way? Because luminance changes aren't important changes most of the time. What really matters is that we detect real objects that we might bump into or that might signal food or a predator. We don't want to have our attention continually drawn to ghosts or shadows. Our visual system is designed by **natural selection** to provide us with information in order to make decisions about what to do in an ever-changing environment.

Evolutionary explanations for failures to reason logically

In 1966, British psychologist Peter Wason published a paper that gave birth to a whole field of psychological research. The paper outlined the results of a simple experiment that purported to show that people often fail to reason logically.

Wason presented participants with four cards, as shown in Figure 7.5, and told them that number cards had letters on the reverse, and letter cards had numbers on the reverse. He then told them a rule, which was "If a card has a vowel on one side, then it has an even number on the other." He then asked his participants which of the cards above, if turned over, was capable of breaking the rule. Before you read on, try the problem yourself.

Stepping through the cards in turn: *A* is a vowel, so if that is turned over and there is an odd number on the reverse, then that breaks the rule; *K,* on the other hand, is a consonant, and therefore the rule doesn't apply (it is about vowels); 4 is an even number, and if there is a vowel on the other side, it conforms to the rule, but if there is a consonant, the rule doesn't apply, as for the *K*. Finally, the 7 card – if that were turned over and there was a vowel on the other side, then that could break the rule because 7 is an odd number. According to logic, therefore, only the *A* and 7 cards could break the rule. Wason found

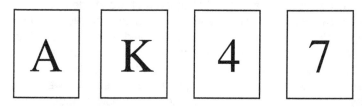

Figure 7.5 Wason's original stimuli were simple: four cards with letters and numbers printed on them.
Source: Adapted from Wason (1966)

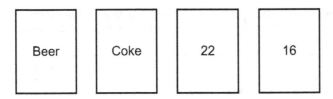

Figure 7.6 Stimuli from the underage drinking task.
Source: Adapted from Griggs and Cox (1982).

that only about 4% of his participants got the correct answer and concluded that this demonstrated that people had difficulty reasoning logically.

A subsequent study showed a very different pattern of response from a very similar problem. Griggs and Cox (1982) told participants that they had the task of determining whether people were breaking the law by drinking under the legal age, which was 19 years. The cards contained drink information on one side and age information on the other (see Figure 7.6). They were asked which cards they needed to turn over to determine whether someone was breaking the law. Try this before you read on.

We can see that if the beer drinker was under the age of 19 they would be breaking the rule. Drinking a coke is no problem as it is non-alcoholic. If the16-year-old were drinking alcohol, they would be breaking the law, but the 22-year-old could not, as they are above the legal age. So, the correct answers are "drinking a beer" and "16 years of age." In fact it was probably unnecessary to tell you that because you probably already realized it: almost instantly. And it was the same for the participants in the original experiment – most of them chose the correct cards.

So what is the difference between these two tasks? One difference is that the original task was making a statement about the truth of a situation (logicians call this an **indicative** task), the later task was about social rules and conventions (called a **deontic** task). There have been many different explanations for why there is a difference, but here we shall focus on one provided by evolutionary psychology.

> **KEY TERM**
>
> **Deontic** and **indicative reasoning** *Deontic reasoning* applies to social rules and obligations. "If you help me move, I will buy you dinner." If you help me move and I don't buy you a meal, then I have cheated on our agreement. *Indicative reasoning* applies to the truth and falsity of a rule, given certain observations. It is therefore not possible to cheat, just to be right or wrong.

Cosmides (1989) considered the free-rider problem discussed in Chapter 5. Recall that one of the problems with cooperation is that it leaves altruists at risk of being exploited by people who take the benefit

Table 7.1 Percentage of correct responses for three different tasks. * The lower figure was when the word *altruist* was used, the higher when the word *selfless* was used.

Task	Percentage of correct responses
Abstract task (letters and numbers)	4
Cheat detection	74
Altruist detection	28–40*

Source: Data from Cosmides and Tooby (1992).

but do not return the favor. Without some means of managing free riders, cooperation among non-kin could never have evolved. Given that cooperation did evolve, Cosmides argues, we must have evolved cognitive mechanisms for detecting those who cheat by obtaining something that they are not entitled to. The under-age drinking problem represents such a situation: people are cheating the system by drinking when they are not permitted to do so. When presented with such a situation, the cheater detection mechanism "fires" and people find it easy to get the correct answer. In indicative tasks, such as the one involving letters and numbers, no cheating is occurring, so participants find it difficult to arrive at the correct answer.

Cosmides and Tooby (1992) compared the performance of participants on the original task versus the cheat-detection task (similar to the under-age drinking problem) and a further task where participants were to detect altruists (who pay the price but don't receive the benefit – the opposite of a cheat) and found that responses are much higher for cheat detection that for all other examples (see Table 7.1).

More recent research has suggested that people are only good at detecting cheats in situations where (1) cheating is possible, (2) cheating is deliberate and (3) cheating benefits the cheat. Otherwise participants are less good at detecting them (Cosmides, Barrett & Tooby, 2010).

Evolution and memory

We frequently curse our memories for failing us. Turning up for the exam at the wrong time, forgetting that important phone number and how many times have you been shocked at how different two people's recollection of the same event could be?

One of the reasons why we curse our memories is that we expect it to record everything in the minutest detail. But this is a vain hope. Not only would we eventually run out of brain space, but it would also take ages to sift through such a colossal pile of trivia. Evolutionists believe our memory was not designed to remember everything; it was designed to remember important things quickly, the better to aid survival. What do we mean by important here? Well there are two aspects to this. The first is the *likelihood* of something happening again. If it looks as if an event is likely to happen again, then it is probably worth remembering more than an event that is unlikely to happen again. The second how *consequential* the event is. Events that are life-threatening or painful are worth remembering more than inconsequential events because they had serious consequences for our ancestors.

To predict the likelihood of an event happening again, memory seems to use the frequency with which the event happened in the past (Anderson & Millson, 1989), past frequency predicts the extent to which something is remembered (Anderson & Schooler, 1991).

One of the factors that can determine how consequential an event is and help decide whether it should be stored or not is emotion. People suffering from posttraumatic stress disorder (PTSD) find themselves reliving traumatic memories, to their obvious distress. Furthermore, people have flashbulb memories (Brown & Kulik, 1977) where they remember – often in minute detail – the events surrounding a socially important even, such as the assassination of President Kennedy, the death of Princess Diana or the atrocities of 911. In all of these instances the emotion – of the individual but also people around them – are used to predict how consequential the event is. So although PTSD might be seen as maladaptive, it could just be a result of the fact that our memories evolved to learn to avoid potentially life threatening situations (Schacter, 2001).

DISCUSS AND DEBATE – MEMORY FOR EVENTS

Think of things that you find particularly memorable. Are there strong emotions associated with these memories? If so were the memories positive of negative? Do you think it is more important, evolutionarily speaking, to remember events that are accompanied by negative or positive emotions? Why?

Finally, recent evidence has suggested that our memory may have evolved to remember things that were of relevance to our ancestors rather than things that are of relevance today. Recall that for most of our evolutionary history humans lived in hunter–gatherer communities on the African grasslands. It may be that this environment has left its mark on our memories (Nairne & Pandeirada, 2008; Weinstein, Bugg & Roediger, 2008). In a typical experimental, condition participants are given the following instructions:

> In this task we would like you to imagine that you are stranded in the grasslands of a foreign land, without any basic survival materials. Over the next few months, you'll need to find steady supplies of food and water and protect yourself from predators.

They are then given a list of 12 words and asked to rate each for its usefulness in their task. In another condition, they are given an almost identical instruction but the word *grasslands* replaced by *city* and the word *predators* replaced by *attackers*. It was found that participants remembered more of the words in the former condition than in the latter. The researchers explain this by suggesting that memory comes pre-prepared to remember things associated with ancestral survival. Rather like a flashbulb memory, the emotions invoked by thinking about an ancestral environment aids memory even when the participants have little direct experience of such an environment.

Summary

From an evolutionary perspective, cognition – be it conscious or unconscious – is largely about an animal answering the question "What shall I do next?" In order to make this decision, other cognitive processes, such as vision, memory and reasoning, need to be employed to ensure that the decision is good, and made quickly. But emotions are also employed as part of this process, and it is to this that we turn next.

Emotion

While evolutionary psychologists have been studying the relationship between cognitive abilities and evolution for some time now, in recent years they have also begun to turn their attention to the relationship between evolution and emotion. If selective pressures have endowed us with a range of cognitive abilities then it makes logical sense to examine the possibility that our emotional responses might also be adaptations (Barrett, 2013; Hess & Thibault, 2009; Nesse, 2009; Shariff & Tracy,

2011). Indeed, as highlighted earlier, without emotions we would find it difficult to make decisions concerning which stimuli we should attend to (and how we should respond). Emotions serve the general function of shifting attention and preparing the body physiologically and cognitively for various threats and opportunities. We will consider in more detail the functions of various emotions after first examining the evidence that the outward manifestation of our emotions – their expression – is the product of evolution. Such a notion harks back to the ideas of Darwin.

Darwin and emotions

Darwin was very interested in emotions – so much so that he devoted an entire book to them in 1872 – *The Expression of the Emotions in Man and Animals*. In *Expression* (the book is regularly referred to by this single-word abbreviation), Darwin suggested that our emotional expressions evolved for communicative purposes and that they arose by natural selection in much the same way as any other trait. For much of the twentieth century social scientists either disagreed with Darwin's views on emotions or more frequently completely ignored them. The prevailing social constructivist view being that emotional expressions are largely learned and specific to a particular culture (Armon-Jones, 1985). Since the 1970s, however, evidence has gradually accumulated that a number of emotional expressions are shared by all cultures, and due to their universality can be seen as adaptations (see Key concepts box 7.2).

KEY CONCEPTS BOX 7.2

Are basic emotional expressions universal?

Based on his travels and correspondence with authorities from around the world, Darwin (1872) came to the conclusion that humans have a number of basic universal emotional expressions. Due, however, to the rise of the blank-slate view of human behavior during the 20th century, it became widely believed that emotional expressions were culturally endowed and hence varied greatly from one culture to another. During the latter years of the 20th century, this view was to be challenged by the findings from a number of field studies. First, American psychologists Ekman and Friesen (1971) reported that tribes people from a remote part of New Guinea were able both to identify Western-specific emotional facial expressions and to reproduce them under similar circumstances.

On the basis of these findings (and those involving a number of other cultures) Ekman and Friesen suggested that Darwin had been right, that we do have a number of basic emotional expressions that served adaptive purposes. Second, German human ethologist Eibl-Eibesfeldt (1973) demonstrated that children who were blind and deaf from birth produced the same emotional expressions under the same emotion-inducing circumstances as sighted children. The fact that they were blind and deaf is evidence that such expressions are innate since they would have been unable to copy them from others. Third, and finally, Eibl-Eibesfeldt was also able to show how individuals from a number of isolated communities not only used the same basic emotional expressions but that they used exactly the same time frame for them. An example of this is the rapid raised eyebrow expression of (pleasant) surprise that we all use when coming across a friend unexpectedly (see Table 7.2). All of these observations suggest basic emotions are the outcome of autonomous processes inherited from our common ancestors.

The emotional expression that have come to be labeled as basic universal include: pleasant surprise, fear, anger, sadness, happiness, disgust and possibly contempt (Ekman & Friesen, 1986). Before we begin to run away with the notion that emotional expressions as simply hardwired and hence genetically coded, we need to bear in mind there are some provisos to this view. First, how often people make use of such expressions varies from one culture to another due to cultural expectations of how appropriate it is to show one's feelings in public. Second, more complex emotions than the basic six or seven, such as expressions of distrust or embarrassment, may be formed from blending the basic emotional expressions, and these, since they involve a degree of learning, may vary somewhat between cultures.

Due to these factors Ekman (1972) has developed what he calls the **neurocultural theory of emotional expression**. That is, basic expressions are universal and used under the same circumstances cross-culturally, but the display rules of each culture can serve to diminish or intensify such expressions and determine when it is appropriate to use them. More complex "secondary" emotional expressions, such as apprehension or embarrassment, certainly involve cultural learning.

Human–animal parallels in emotional expression

Following in Darwin's footsteps, evolutionary psychologists today consider that human emotions (both the internal experience and their external expression) evolved to suit the survival needs of our ancient forager ancestors. They also subscribe to the view that since emotions are adaptations, then because of our common heritage we are likely to share features of them with our primate relatives (a view, once again, first proposed by Darwin). Ever since Dar-

KEY TERM

Anthropomorphism Attributing human-like mental states to animals.

win's day, the notion of sharing emotions with other species has been contentious. Hebb, for example, suggested in 1946 that such an idea is simply **anthropomorphism** (i.e., attributing human attributes to animals). Due to this charge of anthropomorphism, for most of the 20th century biologists and psychologists shied away from the notion of continuity between humans and other species in their emotional responses. Moreover, this reluctance to draw parallels is also due to the not unreasonable suggestion that human emotions are undeniably linked to language (Hess & Thibault, 2009). Despite such misgivings, in recent years there has been a reassessment of the notion of continuity of emotional expression (and experience) between man and ape.

DISCUSS AND DEBATE – HOW CAN WE GUARD AGAINST ANTHROPOMORPHISM?

Darwin regularly discussed animal behavior using terms that are normally reserved for human emotional expression, such as jealousy. Due to this he has been accused of anthropomorphism. If we are to make use of emotional terms when describing animal behavior, what would we need to do in order to avoid this charge?

A small number of studies pointed out human–ape parallels in emotional expression during the 1960s (see Andrew 1963a, 1963b; van Hooff, 1967, 1972) providing support for Darwin. One well-established example of a broadly similar emotional response in humans and apes is the open-mouthed, bared-teeth display that is shown by chimpanzees as a gesture of appeasement, often used by subordinate individuals to more dominant ones. This expression is very similar to the human smile. According to van Hooff (1967, 1972), the human smile evolved as an appeasement signal and has subsequently become a signal of general friendliness. Similarly when the nose is wriggled up and the upper teeth bared in a closed mouth – this is recognised as a threat display in both chimps and humans (see Figure 7.7).

In recent years, researchers have returned to consider the suggestions of Andrew and van Hooff by making use of neurological

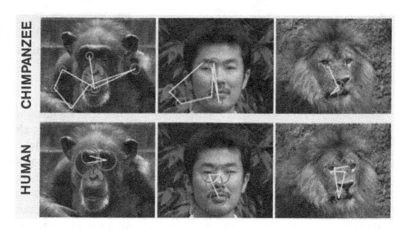

Figure 7.7 "**Gaze plot**" of face-scanning patterns in chimpanzees and humans when viewing faces of chimps, humans and lions.

Source: Kano and Tomonaga (2009). Reproduced with permission.

KEY TERM

Gaze plot A visual representation of how an individual scans an image. The lines represent movements between fixations, and the circles represent how long a particular point is fixated.

KEY TERM

Affective neuroscience Term, attributed to Jaak Panksepp, used to describe the neurological bases of emotional experience and expression.

tools such as scanning and imaging techniques that have recently become available. Neuropsychologist Jaak Panksepp has been a notable supporter of the idea of continuity between apes and humans in emotional states. Panksepp, who coined the term **affective neuroscience** to describe the field that studies the neural mechanisms underlying emotions, has pointed out that we share with them a number of neurological circuits that underlie certain basic emotions, such as rage and fear (Panksepp, 1998, 2004, 2005). This, in itself, suggests such internal states arose from a common ancestor and that there may still be broad similarities. Moreover, humans are very good at predicting the emotional behavior of their pets based on their facial expressions and body language (Fidler, Light & Costall, 1996; Turner & Bateson, 2000). As Panksepp has pointed out, all organisms have to face environmental challenges, and those with well-developed nervous systems, such as birds and mammals, would certainly benefit from making use of such neuronal circuitry to develop emotional responses to overcome such challenges. So far, evidence is emerging that suggests, in addition to high levels of cognitive ability, some of the larger brained ones are putting their neurons to use to analyze emotional information to help them in their decision making.

Looking into the faces of our relatives

One recently developed area of research that has examined human–ape similarities concerns the way that members of each species scans emotional faces of the other. Fumihiro Kano and Masaki Tomonaga, of Kyoto University in Japan, have used eye tracking equipment to uncover both similarities and differences in the scanning patterns of humans and chimps. Making use of five chimpanzee and nine human participants Kano & Tomonaga (2009) found that both humans and chimps initially scanned the eyes of various emotional faces. Interestingly, thereafter, the chimps spent more time focusing on the mouth whereas the humans, having shifted their gaze to the mouth and other parts of the emotional faces, kept returning to view the eyes. This was the case when chimps scan both humans and chimps and vice versa. The fact that both begin by focusing on the eyes suggests that both species use this part of the face as an initial reading of emotional expressions but that humans have developed this reading of expressions from the eyes to a greater degree than chimps. Whether or not scanning patterns differ between different emotional states for chimpanzees is yet to be established.

The function of emotional states

Having established that basic emotional expressions appear to be universal and that they have similarities with our relatives, the great apes, it is now necessary to pose the question – if emotions are adaptations, then what functions might they serve (either today or in our ancestral past)? To put it another way, in what ways might emotions have helped our ancestors survive, gain mates and produce offspring? Curiously, while Darwin emphasized the continuity between humans and other species in *Expression,* he said very little about the function of specific emotions. In recent years, a number of evolutionary psychologists have taken Darwin's ideas further and have begun to suggest ways in which specific emotional states and expressions might provide advantages under the right circumstances (Nesse, 2009, 2014; Fredrickson, 2013).

According to evolutionary psychiatrist Randolph Nesse, research into emotions has been largely concerned with **proximate mechanistic explanations**. That is how they work rather than **ultimate evolutionary explanations** – why they work that way. Recent progress in understanding the underlying neural and psychological mechanisms needs to be mirrored by an understanding of precisely how specific emotional states adjusted individuals in ways that increased their ability to survive and breed in challenging situations during our evolutionary past (Nesse, 2009, 2014; Shariff & Tracy, 2011).

Two stages and two functions model of evolution of emotions

Drawing on the work of Darwin, both Ekman (1992) and Eibl-Eibesfeldt (1989) have independently suggested that the evolution of emotions was a two-step process. First, internal physiological regulation prepared the body for appropriate action to environmental challenges (and selected appropriate cognitive changes – see earlier) and second, the expressions that accompanied such internal changes then came to function for communication with conspecifics at a later stage in evolution. Returning to fear as a basic response, imagine once again the animal that is about to feed but then encounters a dangerous predator. Blood is suddenly redistributed to muscles, and breathing becomes more rapid. Both responses aid escape, but, importantly, these internal changes are *accompanied* by activation of the facial muscles that produce an expression and a cry of fear. Both the cry and the facial response may then be perceived by nearby conspecifics (including, perhaps, relatives). It is clear that these two changes are likely to have been shaped by natural selection since they may serve to aid survival in the individual and their relatives. In this way the emotional response has evolved two functions – survival and communication.

The same two-stage argument can be made for other emotions (both positive and negative), such as disgust. When encountering, for example, rotting food, an animal (including, of course, humans and their ancestors) undergoes physiological changes leading to a loss of appetite (and in extreme cases, feelings of nausea). The accompanying scrunching up of the nose and mouth both reduce air intake and signal to others to avoid the noxious matter. Interestingly, despite our development of culture, negative responses such as fear and disgust remain as stereotypical and exaggerated in our own species as in others less endowed with cerebral development (Shariff & Tracy, 2011). This may be taken as evidence that basic emotions retain their communicative function for our species.

Evolution primes and constrains the development of emotional responses

Today, evolutionists are examining how specific emotional states might have promoted **inclusive fitness** during our ancient past. They have also conducted studies into the degree to which learning is involved in emotional responses. A number of studies have demonstrated how lab-reared monkeys learn to produce appropriate emotional responses to novel stimuli very rapidly by observing such responses in other, more experienced, individuals. An example of this is a study of how lab-reared rhesus monkeys that had previously been unafraid of snakes suddenly become fearful of them after observing wild rhesus monkeys showing fear responses to snakes (Ohman & Mineka, 2001). This suggests that while emotional cries and facial responses may be "hard-wired," the stimuli that are likely to evoke such responses in primates may be learned, but that such learning is primed or constrained by stimuli that would have had fitness consequences in the ancestral past. Interestingly, parallels have been found in humans. In one recent study, parents were required to place their hands into a bag in which an unknown species of animal was contained while being observed by their own children (Creswell, Shildrick & Field, 2011; see also Field & Field, 2013). The experimenters found that when parents produced fearful emotional expressions while placing their hands into the bag, their children picked up on such signals and were thereafter more likely to demonstrate anxious responses.

Positive and negative emotions

There is a tendency to perceive positive emotions as beneficial and negative ones as harmful – unfortunate by-products of the fact that negative ones are induced by bad experiences. According to Nesse, however, this is an illusion we have developed due to their associations with beneficial and harmful circumstances. When threatened, for example, responding with a positive emotion may well be harmful whereas responding with a negative one would most likely prove beneficial. Such a perspective helps us to understand how adaptive having negative emotions within our repertoire must have been and may continue to be today. One evolutionist who has attempted to outline the adaptive value of specific negative emotional states is Paul Ekman. In contrast another evolutionist, Barbara Fredrickson, has attempted to outline the adaptive value of specific positive emotions. We outline each of these in turn.

Ekman (1992, 1999) suggests that fear can be divided into a number of subtypes, each of which has quite specific ways of enhancing inclusive fitness. These include, for example, panic and agoraphobia, which lead to greatly heightened vigilance and flight response. Anger, in contrast leads, to a tendency to attack, which under the appropriate circumstances may have been adaptive during our evolutionary past.

Examples of such adaptive responses might include attacks on competitors or on free riders, which might then signal their nonreciprocation has been detected. Note that as in other species the threat of attack (signaled by appropriate facial expression and body language) is often sufficient to deter others. Sadness is another common negative emotional state. According to Ekman (see also Lazarus, 1991), sadness involves a withdrawal from normal activities and signals that aid is required. Disgust, as discussed above, clearly takes the individual away from noxious stimuli and warns others of the danger. In summary negative emotions – both the internal experience of and the external expression of – are considered to narrow attention, produce the appropriate physiological and behavioral response and simultaneously signal to others the state and intentions of the signaler. In addition to considering the evolution of emotional states, Ekman has also suggested basic human emotional expressions serve quite specific functions (see Table 7.2).

The fact that at least seven basic emotional expressions are recognized universally suggests that Darwin was on the right track back in 1872 and that just like our cognitive abilities, the underlying states that these emotions signal are quite specific adaptations for our species.

KEY TERM

Broaden-and-build theory of positive emotions. A theory developed by Barbara Frederickson that suggests we evolved positive emotions in order to "broaden our mindset." This involves enriching our knowledge base, expanding our social networks and the development of more flexible problem-solving skills.

In contrast to Ekman, Barbara Fredrickson (1998, 2006, 2013) has focused mainly on the adaptive function of positive emotional states. Fredrickson has developed a **broaden-and-build** theory of positive emotions, suggesting that they have evolved to help expand social networks, enrich our knowledge base and build personal resources. In particular, she has considered joy, interest, contentment and love. In the case of joy, this playful emotion enables you to find out about others in a non-threatening way. Joy is often a state that romantic partners experience in each other's company. As such, joy has clear benefits in terms of reproductive output given that it signals such a positive state. With regard to interest, this emotion leads us to explore and increase our knowledge base and flexibility in problem solving – clearly a positive state that can improve survival. Contentment is a state whereby we savor our achievements and, in dwelling on them, are more likely to repeat successful strategies. Moreover, a state of contentment regularly follows childbirth and may help to shift priorities toward the new-born. Finally, love, such as a mother's (and father's) love for her children, clearly has positive consequences for boosting inclusive fitness, as does love of other family members and one's romantic partner.

Table 7.2 Basic human emotions and their proposed adaptive value.

Basic emotion and description	Proposed adaptive value (function)	Facial expression
Anger – eyebrows are pulled down in center, upper and lower lids are pulled up and lips are tightened	Reaction to aversive acts of others. Lets others know you will respond to threats from them and that you will not tolerate free riders.	
Disgust – eyebrows are pulled down, nose is wrinkled up, upper lip pulled up, lower lip loose	Reaction to revolting stimuli (either physical or moral). Helps humans avoid contamination and, later on in evolutionary history, warns others that their actions are morally reprehensible.	
Fear – eyebrows are pulled up and together, upper eyelids are pulled up and the mouth is stretched	Reaction to specific danger. May serve to aid removal from dangerous situations and warn others of such dangers. Panic induced by fear may be inappropriate today but may have prepared individuals psychologically and physiologically for fight/flight/freeze response to aid survival.	
Happiness – muscles around the eye tightened, "crow's feet" wrinkles around the eyes, corner of lips raised diagonally, cheeks raised up	Expression of positive feelings ranging from satisfaction to joy. A goal that people strive for and that is associated with a number of factors that would have boosted inclusive fitness in our past, such as close relationships with others, including family members.	
Sadness – inner corners of eyebrows raised, lip corners pulled down, eyelids loose	An expression associated with loss or difficulty. Signals to others aid is required. Might have allowed sufferer to avoid exertion and helped them to receive support.	
Surprise – brief raising of eyebrows such that eyes are opened widely and the forehead wrinkles. Jaw is relaxed and generally drops down	An expression that arises when encountering an unanticipated situation and in particular seeing an unexpected person. Very brief expression lasting a fraction of a second and is usually followed by other emotions, depending on the stimulus that has caused the surprise. In the case of friends, is followed by a smile that signals alliance.	
Contempt – eyes neutral, corner of just one lip pulled up and back	Reaction to substandard behavior; may warn a member of the social group that they need to improve their behavior.	

Summary

- Cognition in psychology involves mental activity where information is stored, retrieved or transformed. Hence, the cognitive approach treats the mind and brain as an information processor. Prior to the advent of cognitive psychology, the main approach was behaviorism – a field of psychology that treated the mind as a black box. Early proponents of the cognitive approach include Alan Newell, Herbert A. Simon, George A. Miller and Donald Broadbent.

- Computer models, which are programs that are designed to mimic the behavior of humans performing specific tasks, have been developed to help understand human cognitive processes. In recent years, neuroimaging techniques have been developed to tie cognitive processes to physical activity in the brain. Such techniques include PET (positron emission tomography), CAT (computer axial tomography), MRI (magnetic resonance imaging), fMRI (functional magnetic resonance imaging) and MEG (magnetoencephalography).

- One way that evolutionary psychologists have studied the relationship between evolution and cognitive processes is to look for limitations in such processes and relate these to our ancestral past. Cognitive psychologists who are interested in the relationship between evolution and cognition have made use of perceptual illusions in order to understand what our perceptual processes were designed to achieve through natural selection.

- Other cognitively-based evolutionary psychologists have suggested that our regular failures to reason logically are a consequence of the abstract nature of tasks that we have developed today. When Leda Cosmides presented logically just-as-taxing tasks but related them to spotting free riders (a recurrent problem during our evolutionary past), participants were able to complete such tasks when a more abstract version eluded them.

- In the case of memory, evolutionists have suggested that we forget or misremember so much of our everyday lives because our memory systems were designed by natural selection to recall events that would have been consequential during our ancestral past. Highly emotive events tend to be stored and recalled more easily due to their importance for survival and reproduction. Recent studies support the notion that we have better recall for events that would have been of relevance to our forager ancestors.

- Emotions are defined as particular modes that an organism switches to in response to specific internal or external situations in order to promote an appropriate behavioral response. Emotions involve physiological, psychological and usually behavioral responses.

- Darwin was the first person to suggest that our emotions came about due to selective pressure during our evolution. Hence emotions can be seen as adaptations that, like cognitive processes, may have survival and reproductive consequences.

- Darwin suggested that humans have a number of basic emotional expressions (that reflect internal sates). Until the 1970s, this was disputed. Evidence has accumulated over the last 40 years, however, that supports Darwin's view on emotions. First, isolated forager people recognize and make use under the same circumstances of at least six emotional expressions – fear, anger, disgust, happiness, sadness and surprise. Second, children deaf and blind from birth develop these basic emotional expressions – suggesting they are innate. Third, humans recognize emotional expressions on the faces of animals and are able to use these to predict their behavior. Fourth, and finally, our primate relatives show broad similarities with us in their use of emotional expression, such as the appeasement smile and threat display of chimpanzees. Additionally, chimps focus on human emotional faces in much the same way that they focus on the emotional faces of other chimps and in a similar fashion (though not identical) to the way that humans focus on such faces.

- Ekman considers that negative emotions have beneficial consequences in that they aid survival and communicate danger. Positive emotions are considered by Barbara Fredrickson to serve adaptive functions also since they broaden-and-build up relationship and resources.

FURTHER READING

Panksepp, J., & Biven, L. (2010) *The archaeology of mind: Neuroevolutionary origins of human emotion*. New York: W.W. Norton.

Pinker, S. (1997) *How the mind works*. New York: Norton. [The chapter on vision is particularly excellent.]

When things go wrong

Evolution and abnormal psychology

8

What this chapter will teach you

- What is the subject matter of abnormal psychology?
- What is Darwinian medicine (also called *evolutionary medicine* and, when applied to psychiatric problems, *evolutionary psychiatry*)?
- Why humans are susceptible to mental health problems.
- How evolutionary psychologists are beginning to suggest that lifestyle changes may reduce mental health problems.

So far we have considered the relationship between evolution and typical human behavioral responses, such as how we develop, socialize and use our emotions and cognitive abilities to help us understand the world. Given that one in four people living in the West today will experience some form of a mental health issue, however, it is necessary for evolutionary psychologists to consider the field of abnormal psychology. In this chapter we consider why as a species we are so vulnerable to "mental illnesses" and other problems of abnormal behavior, such as autism, within a framework of evolution by **natural selection**. In order to do this, we need to understand why it is that due to constraints on the process, evolution does not lead to perfection. As will be seen, **genes** that have positive effects can also have negative ones (either in themselves or when combined with other genes). We also need to consider the notion that in some cases mental health problems today may be based on adaptations to the past. In order to achieve this, we will introduce the emerging field of **Darwinian medicine,** and in particular the work of evolutionary psychiatrist Randolph Nesse.

What is abnormal psychology?

Abnormal psychology is concerned with the causes and treatments of mental dysfunction and in particular with mental illnesses. The term **psychopathology** is often used as a synonym for abnormal psychology, but is a narrower term since it implies a disease process is occurring. Many psychologists who study abnormal behavior do not, however, consider that pathology underlies such internal states and responses (Bentall, 2003, 2009). Medical practitioners who attempt to treat abnormal behavior and internal states are known as *psychiatrists,* whereas those with training in the psychology underlying abnormal behavior are termed *clinical psychologists.* While many clinical psychologists work alongside psychiatrists in the treatment of patients/clients, there are times when the two approaches come into conflict due to their differing theoretical backgrounds. Put simply, psychiatrists tend to take a medical view (and hence more frequently prescribe drugs treatment) whereas clinical psychologists most often use forms of psychotherapy (talking therapy). In recent, years the two approaches have perhaps moved a little closer since psychiatrists now receive a fair degree of psychological training, and some clinical psychologists (in the USA, for example) are able to prescribe drugs. Academic psychologists have long debated the use of the term *abnormal* when used to describe human behavior that differs from what is normally expected. One problem is that the term *abnormal* can have a number of meanings – such as atypical, unexpected or statistically rare. But there are many examples of behavior fulfilling all of these without it being considered a mental health problem (think of mathematical geniuses and those who are ambidextrous). Also, where exactly do we draw the line between eccentricity and mental illness? Is someone who is convinced their soccer team is the best in the world despite considerable evidence to the contrary (a belief held by one of the authors of this book) deluded or just eccentrically overoptimistic? It would be beyond the scope of this book to review this complex and enduring debate – suffice to say that it is worth being aware that such a debate exists. Fortunately, in relation to abnormal behavior and evolutionary psychology, we can neatly side-step this debate by confining ourselves to examples of mental health issues as recognized by the professional manual that is used world wide in diagnosis – the **DSM-5** (see Key concepts box 8.1).

KEY CONCEPTS BOX 8.1

DSM-5 – A diagnostic bible or just a dictionary?

The *Diagnostic and Statistical Manual of Mental Disorders* (DSM) is the generally accepted worldwide manual used for diagnosing mental health disorders. On May 18, 2013, a new edition (the 5th) was published to mixed reviews. DSM-5 lists all developmental disorders, anxiety and depressive disorders in addition to serious psychotic ones, such as **schizophrenia**. Clinicians often refer to the DSM classification as the "Bible" for the field due to its wide acceptance and use. Such a scheme is controversial, however, due to a number of professional bodies having reservations. The **British Psychological Society** (**BPS**), for example, has suggested that the DSM *medicalizes* natural and normal responses to life experiences. Likewise, the **National Institute for Mental Health** (**NIMH**) has suggested that rather than a "Bible" it is more of a dictionary that creates labels and then defines them. Despite such criticisms, DSM-5 was put together by some of the world's leading psychiatrists and clinical psychologists and does allow people to speak the same language when diagnosing mental health problems cross-culturally.

Darwinian medicine and the relationship between evolution and vulnerability to mental health problems

Traditionally, those interested in abnormal behavior/psychiatry have not embraced the evolutionary perspective. Indeed, psychiatrists and clinical psychologists generally receive little or no training in

Figure 8.1 Evolutionary psychiatrist Randolph Nesse, who is one of the joint founders of Darwinian medicine.

Source: Image copyright © Randolph Nesse. Reproduced with permission.

evolutionary theory. In recent years, however, this has begun to change due largely to the work of one man – Randolph Nesse. Collaborating with leading evolutionist George Williams in the early 1990s, Nesse realized that recent advances in both observational and theoretical work on the relationship between evolution and normal behavior can also be brought to bear on abnormal behavior (Williams & Nesse, 1991; Nesse & Williams, 1995). Over the last 20 years, Nesse has become the major figure in the development of **Darwinian medicine**. (To complicate matters, this is sometimes known as **evolutionary medicine,** and when applied only to psychiatric disorders is often known as **evolutionary psychiatry**.) As we will see, Nesse is not alone in the development of the evolutionary approach to mental health issues, but he has almost certainly been the most influential in moving the field forward through his writings, training sessions for clinicians and public talks.

DISCUSS AND DEBATE – PSYCHIATRY'S RELUCTANCE TO MAKE USE OF EVOLUTIONARY EXPLANATIONS FOR MENTAL ILLNESS

Randolph Nesse has frequently written of his frustration that students undertaking medical training rarely take classes in Darwinian evolution. Why do you think this is the case? (You should be able to suggest at least three different explanations).

Why are we vulnerable to mental health problems?

At this point you may be wondering how evolutionary theory can be brought to bear on understanding mental health problems given that evolution changes organisms in ways that help them to adapt to the environment. It is important to realize that evolutionary psychologists do not assume that all traits are adaptive to today's social and ecological pressures. Current traits can, in as sense, be out of date or, because they are by-products of adaptations, they may leave us vulnerable to health-related problems. Drawing on this knowledge of the limitations of adaptive change Nesse (2012; see also Nesse & Dawkins, 2010) considers that there are six kinds of evolutionary explanations for our vulnerability to ill health:

1   There is a mismatch between aspects of our bodies and the environment of our evolutionary past (to which they were more properly adapted).
2   Pathogens evolve much faster than hosts due to their short generation time and large numbers, which results in costly counter-adaptations to hosts.
3   There are constraints on what natural selection can shape.
4   There are evolutionary trade-offs that keep any trait from being perfected.
5   Some traits that increase reproduction are at the cost of health and longevity.
6   Many protective defenses have negative aspects (e.g., pain and fear).

Note that the first two are based around the fact that evolution is slow, the next two that evolution is constrained and the final two that evolution is misunderstood. As we will see, of these vulnerabilities, all but number 2 have a direct bearing on mental health issues. In the remainder of this chapter will make use of five of Nesse's vulnerabilities list to help explain the existence of a range of mental health issues, such as depression, anxiety, schizophrenia and autism, from an evolutionary perspective. Where appropriate we will also draw on the contributions of other evolutionary-minded researchers in this field, such as Simon Baron-Cohen and Stephen Ilardi. In order to do this we will introduce each of the vulnerabilities with a physical example in order to gain an understanding before considering mental health problems. As will become apparent, the five explanations overlap in that in some cases more than one can be brought to bear on a given disorder.

The mismatch hypothesis

In contrast to our hominin ancestors, many of the most common diseases of modern times are those that are caused by having too many resources rather than too few. These include obesity, diabetes and hypertension (Gluckman & Hanson, 2008). For people living in the developed parts of the globe, these three diseases are largely the result of overindulgence combined with a sedentary lifestyle. Based on studies of current living forager societies, such as the !Kung San of the Kalahari and the Yanomamö of the Amazon basin, when salt, sugar and fat become available there is a tendency to gorge on them. When, for example, hunters of the !Kung San bring back a slaughtered wildebeest, each adult member of the tribe will consume around 2 kg of meat (Lee, 1979). This does not present a health hazard, however, since meat is not eaten in this quantity every day and it follows a vigorous hunt (also, such animals are far leaner than Western cattle). The same argument holds for sugar – with honey and ripe fruit being patchy and unpredictable resources rather than immediately available 24/7. While such current living tribal cultures are by no means identical to our hominin ancestors, it is likely that they share features in common with them – such as hunting and gathering and making decisions as to how they distribute such foodstuffs. And although, as some critics have pointed out, there are gaps in our knowledge as to what exactly life was like for hominins during the Pleistocene, we do at least know what they didn't have. They did not have automobiles, escalators and supermarkets. A sedentary lifestyle was not an option.

If we are adapted to a landscape where access to high quantities of fat, salt and sugar were rare events and a sedentary lifestyle was not available, then, when the opportunity arises to expend less energy and take in extra calories, it is not surprising that people often take advantage of such opportunities today. Ignoring opportunities to feed during the Pleistocene and expending energy unnecessarily would have left an individual less likely to pass on their genes. Hence, raised levels of obesity, diabetes and hypertension can be seen as the outcome of a *mismatch* between the environment in which our ancestors evolved and the one in which many of us live today (Nesse, 2005, 2011; Gluckman & Hanson, 2008).

The mismatch hypothesis – Is your smart phone making you depressed?

While the mismatch explanation is uncontentious when used to explain the serious physical symptoms of an unhealthy lifestyle, it is more controversial when considering mental health problems. Some evolutionary psychologists have suggested, however, that it might be used to explain the serious malady we call *depression*. Clinicians divide depressive disorders up into **major depressive disorder** (also called **unipolar depression**) and **bipolar depression** (Kring et al., 2014; DSM-5). In the case of bipolar depression, there are alternating periods of extremely low, lethargic mood and upbeat, high energy mania (see later). In major depression, by contrast, there are periods of very low mood interspersed by normal mood. Here we consider only major depression, rates for which appear to have grown rapidly since the second World War. Interestingly, the rise in rates of depression has been much steeper in the affluent countries than in poorer ones (WHO, 2008, 2012; Hidaka, 2012). Given the affluent countries are, by definition, materially better off than the less developed countries, this finding appears counterintuitive. It might, however,

make sense if we consider the way the **mismatch hypothesis** has been used to explain this rise. Evolutionary psychologists have developed a number of applications of the mismatch hypothesis to explain this rise in levels of depression, including the influential **social comparison** explanation.

The social comparison explanation, which is a version of the mismatch hypothesis, suggests that because of the rise of social media as our world in effect shrinks, our ability to make social comparisons has become unrealistic. For each of us our sense of self-worth is, in part, determined by comparison with our peers (hence, "social comparison"). During the Pleistocene, making comparisons with others would not have had this devastating effect because while we may not have been the most successful or most attractive person in the tribe, we would not have been subjected to such unrealistic comparisons. Even in the postindustrial age, such idealized images would be rare or non-existent until quite recently. Today, however, given the extraordinarily glamorous and successful images that we are constantly bombarded with from magazines, feature films, TV and the Internet, by comparison we can easily perceive ourselves in an unflattering light. This, then, so the argument goes, reduces our feeling of self-worth and leads to a state of depression for many. Like obesity, the problem would simply not have existed during our evolutionary past. Hence, while depression may have existed back then, it would have been unlikely to have reached the endemic proportions we have witnessed since the development of modern technology, such as the smart phone. It should be noted that this mismatch between the conditions of our ancestral environment and our current conditions might also be applied to other psychological maladies such as anxiety (see later).

The mismatch hypothesis (and in particular the social comparison version of it) appears to make logical sense – but the acid test surely is, can rates of depression (and anxiety) be reduced if we live lives that more closely resemble those of ancestral times? Will you be happier if you put away your smart phone? So far very few attempts have been made to undertake such a project. Intriguingly, one North American clinician and academic has reported positive findings here. Evolutionary clinical psychologist Stephen Ilardi, of the University Kansas, set up a treatment course where clients adopt a lifestyle more akin to our ancient ancestors – called "Therapeutic Lifestyle Change" (TLC). In addition to reducing their social media activities, his clients also spend more time outdoors, increase their physical activity and interact socially more frequently with those around them. Ilardi (Ilardi et al., 2007; Ilardi, 2010) has reported a substantial reduction in levels of depression for those who undertake his 14 week TLC program. Moreover, as Ilardi has documented, people who engage in a lifestyle more similar to our hunter–gatherer past, such as the Amish and the Kaluli of New Guinea, rates of depression are much lower than in the broader American population (Ilardi et al., 2007; Ilardi, 2010).

KEY CONCEPTS BOX 8.2

Is alcoholism a result of a mismatch between current availability and evolutionarily ancient benefits of alcohol?

One interesting application of the mismatch hypothesis comes from Robert Dudley, of the University of Texas, and concerns our thirst for alcohol. Dudley (2002) proposed that our ancestors developed a taste for alcohol long before the evolution of the hominins. The suggestion here, given that ripe fruit is high in sugar and hence calories, frugivorous animals such as our primate ancestors (see Chapter 2) would have gained a major advantages if they were attracted from some distance to fallen overripe fruit that was fermenting. If our early ancestors found the smell and taste of this fermenting fruit pleasurable, then, as a result of the calories it provided, we may have evolved a preference for fruit that contained small amounts of alcohol. The advantages of being attracted to the odor and taste of alcohol may have outweighed any disadvantages since such fruit would, for the most part, have only contained traces of alcohol. This may, however, have left us with the legacy of enjoying the taste and effects of alcohol – which only became problematic once we discovered how to ferment (and distill) large quantities. At this point the mismatch between ancestral conditions and present ones means that an original benefit now has become a great cost for many people who are unable to control their primeval urge to imbibe. While it is difficult to test this theory directly the fact that recent evidence suggests small quantities of alcoholic beverages do have health benefits are conducive to Dudley's hypothesis (Klatsky, 2003).

Constraints on what natural selection can shape

Despite the frequent references by TV naturalists to wild animals being perfectly adapted to their environment, evolution does not lead to perfection. There are limits to the degree to which natural (and sexual) selection can shape traits for a whole series of reasons. One reason is that biological systems are confined by preexisting ones. Unlike the design of a car, where new blueprints can be drawn up from scratch, biological evolution builds on preexisting "design." For example, light entering the eye has to pass through a number of layers before striking the photosensitive retina (a better design would have light hitting the retina first). Another reason is that sex (which, as we have seen is generally considered to be good thing – see Chapter 3) can both break up good gene combinations and bring together bad ones. Related to this is the fact that many behavioral traits are thought to be **polygenic** – which means they are coded for by a number of genes. This means that due to the shuffling of genes that occurs when we reproduce sexually, some people will have too much of a trait and some too little.

Evolutionary constraints and anxiety disorders

Consider our anxiety responses. Anxiety no doubt helped our ancestors to survive (those incapable of experiencing this state would have made a good meal for any passing predator). Given, however, it is almost certainly a polygenic trait, this means that some people experience too much anxiety and some too little. Both can be problematic – imagine if you quit your job without having another to go to or if you were too afraid to leave home (agoraphobia). At either end of the distribution of proneness to anxiety the level of response is problematic. Hence, due to three limitations of what natural selection can achieve (we can't jettison the old genes, limitations of sexual reproduction and of polygenic traits) people who find themselves at the extremely anxious end of the distribution are deemed to have an **anxiety disorder**. In addition to phobias, other common anxiety disorders include **separation anxiety disorder** (where people cannot bare to be left alone) and **social anxiety disorder** (where people find speaking or performing in front of others unduly stressful). All of these problems are classified by DSM-5 as disorders if their levels are such that sufferers are debilitated by them. Note that the problem of a high prevalence rate of anxiety disorders today (one study found that across six European countries the average prevalence rate of anxiety disorders was 16.6%; Somers, Goldner, Waraich & Hsu, 2006) may be exacerbated by features of the modern environment that did not exist in our ancestral past, such as crowded cities and fear-inducing news stories. This means that both the mismatch and the genetic constraints explanations can be brought to bear on why anxiety disorders are so common today. Constraints on evolutionary processes might also help to explain a number of other problems, such as depression and schizophrenia.

Design "Trade-offs" – Are there compensatory advantages to mental health problems?

The notion of design "trade-offs" means that positive traits can evolve that also lead to negative traits provided the former ones outweigh the latter. Remember, as outlined above, natural selection does not perfect but selects for features that boost inclusive fitness (or would have done in our ancestral past). Many traits are trade-off compromises which, on balance, are supported by selection processes. If a design engineer were to consider any organism, they would soon find that, despite 3.5 billion years of evolution, they could quite easily make improvements. Consider our bones – if we apply severe pressure to them there comes a point when they will fracture. We might have evolved stronger, thicker bones that would only fracture when twice the pressure had been applied. This sounds like an improvement – but the extra weight would have slowed us down (speed probably saved many ancestors' lives in times when predators were rife). So, as a result of this degree of trade-off of strength for lightness, bones, on balance, work pretty well.

In addition to physical ones, we can also make this trade-off argument for psychological traits. Above we considered major or unipolar depression. Some evolutionists have suggested that a second form of depression – bipolar depression (or bipolar affective disorder, as it is more correctly known) may be the result of an evolutionary design trade-off.

KEY CONCEPTS BOX 8.3

Heterozygous advantage – A special case of the trade-off explanation

In genetics it has long been known that a specific **allele** (a version of a gene at a particular locus – see Chapter 2) can be maintained in a population when there are benefits associated with having two different versions of the gene. This is known as **heterozygous advantage** and means that when an individual has different alleles of the same gene at the same locus on homologous chromosomes, they may have a fitness benefit when compared to other alternative genotypes. In fact, this heterozygous advantage may allow a **homozygous** condition to remain in the population that actually reduces fitness because of the advantage for some offspring outweighs the disadvantages for others. The best way to illustrate this is by considering a condition known as sickle cell anemia. (If you have studied biology, you might be aware of this condition – if you haven't, it might be worth reviewing basic Mendelian genetics in Chapter 2.)

> **KEY TERM**
>
> **Heterozygous advantage** A situation where a heterozygous genotype provides fitness advantages when compared with homozygous alternatives.

Sickle cell anemia (SCA) is an inherited disorder where the sufferer's red blood cells are sickle or crescent-shaped rather than the normal disc shape. Because of their abnormal shape they do not covey oxygen as well as normal red blood cells and tend to block blood flow in capillaries. The disease is caused by having two different alleles that are normally labelled A and S. Those with the genotype AA have normal red blood cells, whereas those with AS, as well as having normal red blood cells also have a degree of immunity from malaria. This means that individuals with the AS genotype have enhanced fitness compared to the rest of the population in areas where malaria is common. Hence, people who are AS produce more surviving offspring than either AA (normal) or SS (show SCA) genotype. If you recall using a **Punnett square** (see Chapter 2) we can represent the most likely outcomes from a pairing of two AS individuals in the following way:

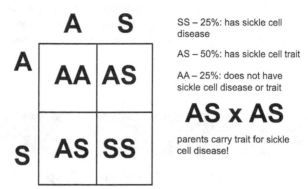

SS – 25%: has sickle cell disease

AS – 50%: has sickle cell trait

AA – 25%: does not have sickle cell disease or trait

AS x AS

parents carry trait for sickle cell disease!

Figure 8.2a Heterozygous advantage – A special case of the trade-off explanation.

If the Punnett square is difficult to visualize, then this diagram should help (note that "carrier" means they carry the disease but have some immunity for malaria):

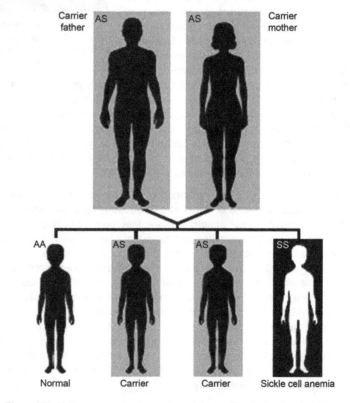

Figure 8.2b Heterozygous advantage – A special case of the trade-off explanation.

Hence because 50% of the offspring have an advantage (some immunity to malaria) and 25% are normal, the fact that 25% have the condition leaves the parents better off in terms of the overall fitness. This is a special case of the trade-off explanation because the parents are, in effect, trading off one sick offspring for one normal one and two with superior fitness. It is entirely possible that heterozygous advantage may help to explain the existence of a number of psychological illnesses in humans in the future – suggesting that we should be studying the siblings of sufferers to determine whether they have some trait that gives them superior fitness.

Is the notion of the mad genius due to a trade-off?

The notion of the mad genius is a widespread stereotype. Think of musicians fervently composing all day and night taking little sleep and ignoring their friends and families, or poets talking to themselves and scribbling away in the middle of the night. Such behavior is considered at the least eccentric and at worst mad. Like many stereotypes there is some truth in this portrayal of the mad genius artist. Over the last 30 years, a number of studies have provided evidence that highly successful individuals in the creative arts have a much greater chance of being diagnosed with bipolar affective disorder (or a related mental illness) than the rest of the population (Jamison, 1995, 2011). Examples of such individuals (some of whom have been "diagnosed" retrospectively) include Vincent van Gogh, Edgar Allen Poe, William Blake, Sylvia Plath, Tennessee Williams and Robert Schumann. In fact, a study by Nancy Andreasen (1987) uncovered the astonishing prevalence rate of bipolar (and sometimes unipolar) depression of 80% for the faculty of a prestigious Iowa Writers' workshop. In a similar fashion, in her book *Touched with Fire: Manic Depressive Illness and Artistic Temperament*, Kay Jamison found that while bipolar illness is found in 1% of the general population, in the artistically talented people she studied the rate was 38%. She has also documented the fact that during their manic phase, the speech of sufferers contains three times as much alliteration and idiosyncratic, rhyming words than that of unaffected individuals (Jamison, 1993). This suggests that creativity is strongly associated with such manic periods.

You may be thinking at this point that creativity can bring rich rewards today, given that we all pay to see and hear great artists, but how would that have helped in our evolutionary past? It has been suggested that the creativity that finds an outlet in music and fine art today may have been utilized to solve problems of a challenging savannah environment in the past (Andreasen, 1987; Jamison, 1993, 1995, 2011). Imagine, for example, the sudden creative insight that leads to a new and more effective arrowhead or a new method of producing fire. Such creativity might have helped our ancestors and their families to survive. Perhaps for reasons yet unknown the extreme depression of the low phases is the price such artists have to pay for their brilliant creativity during the manic phase?

This scenario may sound a little imaginative, and is by no means proven. Given, however, it is relatively common (1% is a much higher rate than would be expected were it to be maintained by random genetic mutation) and given there is clear evidence of a genetic involvement (Gershon, Martinez, Goldin & Gejman, 1990 – see Key concepts box 8.4), then the fact that natural selection has not removed it from our species suggests it does requires some sort of **ultimate explanation.** The notion that bipolar affective disorder is maintained in our species because it provides compensatory advantages that outweigh the disadvantages (i.e., an evolutionary trade-off) is certainly one possibility. Since the trade-off argument suggests that the condition is related to, or is a part of, an adaptation, then we might predict that, unlike unipolar depression (which you will recall was explained by the mismatch hypothesis), bipolar depression will have a higher rate of **heritability** (see Chapter 2). Currently, there is evidence that this is the case – see Key concepts box 8.4.

KEY CONCEPTS BOX 8.4

Is depression inherited?

It is commonly observed that depression runs in families. Might this suggest that it is hereditary? It might – but we have to bear in mind that many non-genetic features run in families, such as poverty (and recipes for fudge cake). The finding that the more closely related you are to a sufferer of depression, the greater are your chances of developing it yourself is at least suggestive that heredity factors (genes) are involved. Interestingly, while having a relative with unipolar (major) depression somewhat increases the chances of succumbing to the disorder yourself, having a relative with bipolar depression greatly increases the chances of developing the disorder (Gershon et al., 1990). How has this been determined? One reason we know this is due to studies of different type of twins that compare **concordance** rates. Concordance, when used in genetics, refers

KEY TERM

Concordance rate Term used in genetics as a measure of the likelihood of an individual for a pair (such as twins) having a specific trait if the other individual in the pair has that trait.

to the probability that a pair of individuals will both have a trait given that one of them has it. The table below compares the concordance rates of two types of twins – **monozygotic** and **dizygotic** (produced by one and two fertilized eggs, respectively – see Chapter 6) for these two forms of depression. In this case the probability is provided as a percentage. Note that in the following table, monozygotic twins have a higher concordance rate than dizygotic twins. Because monozygotic twins share all of their genes and monozygotic twins share only half of their genes (by common descent), the fact that both forms of depression show a higher concordance rate for this type of twin suggests that both forms of depression have a degree of heritability. The acid test, for the notion that bipolar depression has a higher rate of heritability than unipolar is, however, the increase in concordance rates for bipolar when compared with unipolar depression for both mono- and dizygotic twins. The fact that bipolar is more heritable (see Chapter 2) than unipolar depression might also be taken as evidence that it is maintained in the population as an adaptation (or a side effect of an adaptation) whereas unipolar is less likely to be explained in this way.

Concordance rates for two forms of depression in twin studies

Type of twin	Bipolar	Unipolar
Monozygotic (100% of genes shared)	70%	40%
Dizygotic (50% of genes shared)	34%	11%

DISCUSS AND DEBATE – CRITICISM OF TWIN STUDIES

The use of twins in studies of concordance rates has been criticized by some researchers (see Key concepts box 8.4). Make a list of the advantages and disadvantages of such studies when it comes to understanding mental health problems, such as uni- and bipolar depression.

Schizophrenia – Another form of mad genius, or an over complicated brain?

During the manic phase of the illness the symptoms of bipolar affective disorder overlap with those of what is arguably the most serious and disturbing mental illness – schizophrenia. The nature of

KEY TERMS

Schizophrenia A serious mental illness whereby the sufferer exhibits a number of bizarre behavioral symptoms, such as hallucinations, delusions, disorganized thoughts and in some cases, paranoia. According to DSM-5, symptoms must be present for 6 months, during which they must be active for at least 1 month.

Dissociative identity disorder Disruption of identity characterized by two or more distinct personalities or the experience of being possessed. Used to be known as *multiple personality disorder.*

Social brain hypothesis The idea that schizophrenia is a result of having evolved such a complex brain to deal with social complexity that in some cases neurodevelopmental processes do not proceed properly, leading to psychiatric problems.

schizophrenia has been debated ever since the label was first coined by Eugen Bleuler in 1908, but the symptoms, which are described in DSM-5, are well recognized by psychiatrists. Schizophrenia literally means "split mind." This is different from **dissociative identity disorder** (known colloquially as *split personality*), where the sufferer considers themselves to have two or more personalities, since the split in schizophrenia is from reality. (Note that when someone, who is in two minds about something, says they "feel schizophrenic about that," they are using the term inappropriately.) This split from reality means that sufferers of this debilitating illness are likely to have disordered thoughts and feelings. Additionally, they may hear voices (very common) or have visual hallucinations (less common) and grandiose false beliefs (they may think they are some historical figure – although not necessarily Napoleon, as is commonly conceived). They may also be paranoid, believing their loved ones are conspiring against them – often because the voices they hear tell them this.

As for bipolar depression, the rate of occurrence of schizophrenia is also around 1% of the population, and this has been found cross-culturally, suggesting it is a universal phenomenon. Again, like bipolar depression, there is clear evidence that it is heritable (Kring et al., 2014). If a first-degree relative has schizophrenia, then your chances of developing it rise from 1 in 100 to 1 in 10. Also, if a child has two parents with schizophrenia, they have 50–50 chance of developing the illness. Studies have demonstrated that inherited factors are not sufficient for the development of the illness, and this has to be combined with some sort of environmental stressor. Hence, the genes that sufferers inherit make them more prone to developing the illness. Historically, people with schizophrenia were often seen as having been possessed by evil spirits, and until the 20th century sufferers were generally dealt with harshly by being incarcerated in lunatic asylums, such as Bethlehem Hospital, from which we derive the term *bedlam* (see Figure 8.3).

Much ink has been spilled on the relationship between evolution and schizophrenia – too much to be examined in detail here (see the Further Reading section). Two will be discussed briefly. One influential explanation that has been relatively well received is called the **social brain hypothesis**. The social brain hypothesis, which was introduced by

Figure 8.3 An artist's (Hogarth – Rake's progress) impression of what it was like for the "patients" of "Bedlam."

Source: William Hogarth [Public domain], via Wikimedia Commons.

Jonathan Burns (2007), suggests that the evolution of the hugely complex human brain was driven to its current size due to the advantages of processing information about social cognition (i.e., interpreting and acting on the thoughts motives and feelings of others – see Chapter 2). In order to complete this task, the human brain became so complicated and required so many genes (and the appropriate environmental input) for appropriate neurodevelopment to take place. This means that any slight deviations from the normal path of development can lead to the illness. Because the illness consists largely of overinterpreting the social behavior of others then getting this badly wrong, this does sound like quite a good description of the illness. Another suggestion is that the genes that are involved in schizophrenia are related to language development, and sometimes this goes badly wrong. Crespi, Summers and Dorus (2007) identified 28 genes that they believe are related to the neurodevelopment required for language but also are involved in susceptibility to schizophrenia. Crespi and his coworkers suggest that various combinations of these genes promote creativity, and particularly linguistic skill (hence, this may overlap with bipolar depression). Having all or most of them, however, results in a propensity to develop schizophrenia. This means the genes that can lead to schizophrenia, given a stressful environment, are maintained in the population due to the advantages they provide in terms of creativity and linguistic skill. The genes are therefore seen as having been selected for due to their advantages, with some people who have the wrong combination developing schizophrenia as a by-product of the selection. Hence, we might call this the **language by-product hypothesis**.

These two ultimate (evolutionary) explanations for schizophrenia overlap in that they both suggest it is the outcome of having such a complicated brain. They do have some differences however. The social brain hypothesis can be seen as a trade-off explanation since it is the ability to understand the intricacies of others that led to the genes being selected for this ability but also left the brain open to disruption. This is, however, a different trade-off explanation since the sufferer does not gain an advantage but rather the genes, on balance, for such a brain development provide advantages, perhaps, for the majority of relatives who do not develop the illness. (You might recall from Chapter 5 that according to kin selection theory we can pass on copies of our genes indirectly through relatives other than our offspring.) For the language by-product hypothesis, there are clear advantages for the individual, provided they have periods of creativity prior to succumbing to the illness (hence, as in bipolar depression, the compensatory advantage trade-off of the "mad genius" argument is also being made). Or, as with the social brain hypothesis, the genes may be kept in the population because they endow advantages to the relatives of the sufferer. Currently, the view that schizophrenia is maintained in the population because of these advantages is not widely accepted (Nesse, 2011, see also Brüne, 2004).

Traits that increase reproduction are at the cost of health and longevity

People often assume that evolution by natural (and sexual) selection leads to organisms living a longer life. This can be the case since selection clearly leads to the development of antipredator adaptations in many species. But we should not lose sight of the fact that selection boosts inclusive fitness, which in most cases means increasing reproductive success rather than necessarily increasing longevity. In the case of males, as we saw in Chapter 3, the elaborate features and risky behavior many exhibit during courtship can both damage their health and even shorten their lives. Also, males of many species engage in aggressive competitive encounters that can also shorten their lives. In our own species, the reason men have a shorter life expectancy than women is due directly to their differing roles in the game of reproduction (the difference is around 4 years in the UK and USA, for example; WHO, 2013). Because females invest more in the production of offspring, males compete for access to them. This leads to a higher level of risk-taking behavior and a higher rate of mortality throughout life for males. Mortality due to higher risk taking, however, explains only part of the difference in life expectancy for men. Testosterone, which helps males to build greater upper-body strength (driven both by female choice and male/male competition) also suppresses the immune system and leads to prostate cancer for a large number of men (a 2011 report demonstrated that prostate cancer is the second most frequently diagnosed cancer in men worldwide; Jemal, Bray, Center, Ferlay, Ward & Forman, 2011), while a 2007 study found that 80% of men will develop prostate cancer if they live to 80 years of age (Bostwick & Cheng, 2014). Males could live as long as females if they had this source of testosterone removed at an early age (as evidenced by the 19-year increased lifespan of castrated Korean eunuchs; Min, Lee & Park, 2012). But, of course, they would not be able to pass on their genes. In a very real sense, increased reproductive success for males can and often does shorten their lives.

DISCUSS AND DEBATE – WHY DO MORE MEN THAN WOMEN DIE OF CANCER?

Across the world, more men than women die each year of cancer. Suggest three reason why this is the case. How many of these are proximate ("here-and-now") explanations and how many ultimate (evolutionary) ones? How might you determine which explanation is correct? Can two or more explanations be correct?

The relationship between testosterone, the extreme male brain and autism

The fact that men succumb to more life-shortening physical health problems than women is indisputable. Moreover, there is clear evidence that testosterone is involved in this sex difference in life expectancy. The

KEY TERMS

Autism spectrum disorder (ASD) A range (spectrum) of developmental disorders that involves social and communication deficits. ASD sufferers may have difficulties forming friendships and may find disturbing changes to routines or to the environment.

Extreme male brain hypothesis The notion that individuals with autism have an overly male-type brain that is better able to systemize rather than empathize.

Empathizing The ability to read and respond to the emotional state of others. Believed by some to be better developed in the average female than the average male. Can be scored as an *empathizing quotient.*

Systemizing The ability to apply rule-governed skills to processing information. Believed by some to be better developed in the average male than the average female. Can be scored as a *systemizing quotient.*

question for Darwinian medicine/evolutionary psychiatry is, however, is it possible that we can extend this argument to cover disorders that come under abnormal psychology? One, on the face of it, contentious idea which was proposed by Simon Baron-Cohen (2002) is that people with the developmental disorder of **autism (autism spectrum disorder –** DSM-5) have an **extreme male brain,** and that this is related to elevated levels of circulating testosterone during fetal development. Baron-Cohen and his collaborators proposed that, in addition to testosterone playing a role in male competitive and reproductive behavior, it also shifts the developing brain along the path towards **systemizing** rather than **empathizing**. Systemizing, according to Baron-Cohen, involves typical male-type skills, such as perceiving entities as rule-governed systems (as is the case in, say, engineering). In contrast, empathizing involves more female typical skills, such as the ability to understand the feelings of others and respond with the appropriate emotions. The argument is not that each sex can only do one of these things, but that males, on average, are better at the former and females, on average, are better at the latter. Some evolutionary psychologists have suggested this difference is due to the differing challenges that our male and female ancestors faced and hence the differing adaptations they developed. Ancestral men may have evolved a superior ability to systemize since this skill would have helped them to fashion better tools and develop superior spatial navigational skills necessary for hunting. Ancestral women, for their part, would be under selection pressure to develop empathizing skills in order to understand the needs of children (Baron-Cohen, 2003; Geary, 1998).

In the case of autism, where the individual has problems understanding emotional states of others (but is often good with mechanical abilities), the hypothesis here is that the advantages that testosterone generally provides for most males reaches a tipping point for a minority. This then leads to a brain that is good at systemizing at the expense of any ability to empathize – the extreme male brain theory of autism. This notion of an extreme male brain also makes sense given that autism is seen as a spectrum disorder today whereby an individual can be plotted on a continuum from very mild to severe. Perhaps most males have a degree of autistic-like ways of dealing with the world relative to females.

This extreme male brain hypothesis has been heavily criticized, not least by feminists who see it as supporting biological determinants rather than cultural ones to explain gender differences (Nash & Grossi, 2007). Recent research findings would appear, however, to provide support of this hypothesis. One important finding in particular has made use of amniocentesis, which allows researchers to determine levels of fetal testosterone (FT) and later relate these to scores on tasks involving systemizing and empathizing. Astonishingly, Baron-Cohen and his coworkers were able to demonstrate that FT levels are positively correlated with scores on a **systemizing quotient** test whereas scores on an **empathy quotient** test were negatively corre-lated with FT levels (Baron-Cohen & Wheelwright, 2004). This can be taken as evidence that high levels of testosterone during fetal devel-opment can boost systemizing systems at the expense of empathizing ones. Moreover, it has also been found that elevated levels of FT are associated with the development of autism (Ruta et al., 2011). Finally, there is now evidence that specific genes that are related to the manufacture of testosterone are associated with autism (Chakrabarti et al., 2009). As Baron-Cohen has pointed out, these findings are consistent with the fact that males are diagnosed with autism four times as often as females (Baron-Cohen, 2003, 2012; Baron-Cohen et al., 2014).

Why do some females have autism?

If the extreme male brain hypothesis is correct, then this raises the question why do some females develop autism? The fact that some girls develop autism is curious but is by no means a death knell to the extreme male brain hypothesis. While female embryos do not normally have the same levels of FT as males embryos, in cases where females embryos do have elevated levels of FT in the womb they have been shown also to have lower scores on an empathy quotient test and higher scores on the systemizing quotient test (Auyeung et al., 2009, Auyeung, Taylor, Hackett & Baron-Cohen, 2010).

Today, the extreme male brain hypothesis for autism is continuing to gain scientific support as Baron-Cohen and his coworkers have uncov-ered specific differences in areas of the brain that they consider are related both to systemizing and to autism (Lombardo et al., 2012; Lai et al., 2012; Ruigrok et al., 2014). The case is far from proven, however, as many dispute their interpretation of such findings. Finally, another impor-tant point to note is that if autism is a result of selection for relatively high levels of testosterone in the organization of the typical male brain, then we might also think of it as the result of an evolutionary trade-off (see earlier).

Many protective defenses have negative aspects

Many unpleasant symptoms of physical illness are caused by our own body's defenses against pathogens. These include pain, high temperature, coughing, nausea and vomiting – all of which have been shaped by natural selection to expel or overcome the invading foreign body. Hence, our protective defenses are generally unpleasant for us because they are unpleasant for the bugs (vomiting and coughing are attempts to expel the bug, and a raised body temperature also does so since most pathogens can only survive within a narrow temperature range). Many clinicians see these symptoms as abnormalities rather than adaptations – a state of affairs Nesse calls "the clinician's illusion" (Nesse, 2006). The clinician's illusion may also be true for psychiatry.

Enduring negative emotional states – A necessary evil?

Is it possible that some mental health problems might be in part a result of our body's defense mechanisms? As we saw in Chapter 7, emotional states evolved to adjust behavior in ways that aided our ancestors' abilities to deal with environmental and social challenges. Although there is a tendency to consider positive emotions as normal and negative ones as maladaptive, as Nesse (2006) has pointed out, both are equally adaptive in that they would have had survival and reproductive implications.

Negative emotional states, such as fear, are necessarily unpleasant in order to shift priorities to deal with the challenge they signal (see Chapter 7). Enduring unpleasant negative emotional states become defined as psychiatric problems – such as forms of depressive and anxiety disorders (see above). One evolutionary-based idea is that the negative emotions would generally have been relatively transient during our ancestral past as the problems would have been resolved quite quickly (e.g., the fight-or-flight response when dealing with a dangerous animal or unknown tribe). It is only under the conditions that we have created in recent years that such problems endure (e.g., fight or flight will not solve the problems of a 9-to-5 bullying line manager). Note this explanation is a form of the mismatch hypothesis discussed earlier. Another explanation for enduring negative emotional problems is that, as suggested for autism and depression, given the utility of having them in our repertoire, some people at the extremes of the distribution simply have too much of them. Finally, we can add to this list of potential ultimate explanations the fact that, although negative emotional states, such as anxiety, may have evolved to be unpleasant in order to drive us to change our circumstances, given today's pressures, we might not be able to change the circumstances. Hence, we may have enduring unpleasant internal states that simply did not occur in our evolutionary past. In this way, anxiety and depressive disorders may be due to a combination of mismatch, trade-off and the fact that protective defenses have negative aspects.

Darwinian medicine – A step forward?

As we have seen, the hypotheses that have been proposed to help understand our vulnerability to mental health disorders are not separate and discrete but rather overlapping explanations, a number of which can often be brought to bear on the same disorder. It is unlikely that any mental health disorder will be explained by a single cause. The notion, however, that high rates of mental illness are related to the rapid changes in lifestyle that our species has undergone since leaving the open savannah is a powerful one. Additionally recent changes in lifestyle for those living in the wealthiest nations may have taken us even further away from the conditions under which our species evolved, leading perhaps, in turn, to rapid increases in psychiatric problems. Given such a rapid rise, Darwinian medicine is arguably the area where evolutionary psychology can have the greatest impact within the field of evolutionary psychology. It is, however, also the area that is arguably the most contentious (Workman & Reader, 2014). Clearly, many of the explanations that have been developed by evolutionary psychologists are speculative. Initial studies of lifestyle changes that bring us closer to the environment of our ancestors on mental health have had some positive results (Nesse, 2005, 2011). Many, however, remain to be convinced.

Summary

- Darwinian medicine is concerned with the relationship between evolution and human vulnerabilities to illness. It is also called *evolutionary medicine,* and when restricted to mental health issues – *evolutionary psychiatry.*
- Abnormal psychology deals with all aspects of abnormal behavior and internal states. Hence, Darwinian medicine and evolutionary psychiatry are concerned with understanding illness and abnormal behavior from an ultimate perspective.

- Evolutionary psychiatrist Randolph Nesse has suggested there are six kinds of evolutionary explanation for our species vulnerability to illness:
  1. There is a mismatch between our current environment and our evolutionary past.
  2. Pathogens evolve much faster than hosts resulting in costly counteradaptations.
  3. There are constraints on what natural selection can shape.
  4. There are evolutionary trade-offs that keep any trait from being perfected.
  5. Traits that increase reproduction are at the cost of health and longevity.
  6. Many protective defenses have negative aspects.
- It is argued that all but the explanation for our vulnerability to pathogens can be used to understand mental health problems.
- A version of the mismatch hypothesis – the social complexity hypothesis – explains major depression as a result of loss of self-worth due to unrealistic comparisons people make with glamorous images from the media.
- The constraints on natural selection to shape adaptations hypothesis has been used to explain anxiety disorders, such as separation anxiety disorder (anxiety occurring when left alone). In this case high levels of anxiety are associated with being at the extremes of a normal distribution.
- One explanation for bipolar affective disorder, where there are period of high and low mood, is that the creative drive we see during the manic phase of the illness provides compensatory advantages which outweigh its debilitating aspects.
- A number of ultimate explanations have also been proposed to explain the occurrence of another serious mental illness – schizophrenia. One of these is the social brain hypothesis, which suggests a brain that is designed to deal with social complexity can, under the wrong circumstances, develop to overinterpret the motives and behavior of others leading to paranoia and other symptoms such as hearing voices and having disorganized thoughts. Another ultimate explanation suggests that the genes that have been selected for to aid the development of language and creativity, when found in the wrong combinations, can also lead to these symptoms.

- Autism spectrum disorder, where the sufferer has difficulties with communication and areas of social life, has been explained as being the outcome of the development of an extreme male brain. That is a brain that is good at systemizing (perceiving entities as rule-governed systems) at the expense of empathizing (being able to pick up on the feelings of others). It is suggested that this disorder is the outcome of the hormone testosterone, which aids male reproductive behavior, being badly modulated during fetal brain development.
- Enduring negative emotional states that are seen as mental illnesses, such as depression and anxiety, may be due, in part, to the fact that the ability to experience unpleasant states was necessary for the survival of our ancestors. It is suggested that these negative states would have been more transitory survival mechanisms during our ancient past whereas in the current environment they may be prolonged to the point where today they are classified as mental illnesses.

FURTHER READING

Brüne, M. (2008) *Textbook of evolutionary psychiatry: The origins of psychopathology*. Oxford: Oxford University Press.

Gluckman, P., & Hanson, M. (2008) *Mismatch: The lifestyle diseases time-bomb*. Oxford: Oxford University Press.

Kring, A. M., Johnson, S. L., Davison, G. C. & Neale, J. M. (2014) *Abnormal psychology: DSM-5 update* (12th ed.). New York: Wiley.

Culture vultures
Evolution and culture

9

What this chapter will teach you

- A definition of what culture is, and how it has been studied and thought about historically.
- A consideration of the focus on cultural differences at the expense of cultural similarities.
- We then discuss Brown's work on cultural universals and discuss their implications for evolution.
- The psychological mechanisms of culture are discussed.
- The importance of specialization in the transition from hunter-gatherer societies to large-scale civilizations.

What is culture?

Culture is an ambiguous word. In common language it is usually taken as referring to something akin to the high arts, such as music (usually classical), painting, poetry, theater and sculpture and usually has positive connotations: being "cultured" is generally seen as a good thing.

This meaning of the word *culture* is closest to its original usage. It was first used by the ancient Roman orator Cicero (106–43 BCE), the word coming from the same root as "cultivate." In the same way that a farmer cultivates plants to improve yield or flavor, so it was recommended that humans cultivate their habits by attempting to better themselves, and art was certainly seen as one way of doing this (it is no accident that agriculture and horticulture contain the same word).

To a social scientist, however, culture refers to the totality of practices and artefacts that are embodied by a particular group of people: everything from the language you speak to the way you wear your hair; from customs surrounding marriage to death rituals and rites of passage.

Culture is often contrasted with biology. Just as we wouldn't consider opera as part of biology, we would not consider digestion, mitosis or synaptic transmission as part of our cultural heritage. In fact, many researchers explicitly define culture as something that is learned rather than something that is innate. For example, the anthropologist E. A. Hoebel (1966) defined culture as "the integrated system of learned behavior patterns which are characteristic of the members of a society and which are not the result of biological inheritance" (p. 5). Many other social scientists followed this definition; for example, Carter and Qureshi (1995) similarly defined culture as "a learned system of meaning and behavior that is passed from one generation to the next" (p. 241).

The study of culture in psychology is bound up with study of cultural differences. Psychologists now see the way that different peoples' behave as partly influenced by their culture. But it wasn't always like this, and it is instructive to examine a little bit of history in order to understand why things are the way they are now.

A brief history of culture

The study of culture in its own right began in earnest in the 18th century, when Western explorers such as Captain James Cook and Wilhelm von Humboldt encountered cultures that were very different from their own. Taking their guide from the prevailing methodology of naturalists at the time, theorists developed taxonomies of cultures. For example, a culture might be categorized as "civilized," "primitive" or "tribal." Furthermore, the reasons for these perceived differences in cultural complexity were seen by many as being the result of biological differences. The otherwise great German biologist Ernst Haeckel – and he was by no means alone in expressing this view – stated that "natural men are closer to the higher vertebrates than to highly civilized Europeans" (Haeckel, 1904, p. 450). So, in contrast to the definitions expressed above, early ideas about cultural differences were attributed to biological differences.

This view changed in the late 19th and early 20th centuries when anthropologists from the **cultural relativist school** established by Franz Boas (1852–1942) argued that rather than culture being the result of the people, people should be seen as the product of their culture. This view led one of Boas's students, Alfred Kroeber, to argue

that culture was **superorganic**. The word *superorganic* refers to the treatment of culture as if it were an organism in its own right, with its own autonomy and the ability to shape people in any way it sees fit. Sociologist Ellsworth Faris summed up this position when he wrote in 1927 that "instincts do not create customs; customs create instincts, for the putative instincts of human beings are always learned, never native" (quoted in Degler, 1991, p.161).

This is a reversal of the previous view, which saw cultural differences as the result of biological differences: Culture is seen as an autonomous superorganism, as if it is a powerful and autonomous living thing.

Boas's most famous student was probably Margaret Mead. As a young social anthropologist Mead published *Coming of Age in Samoa,* in which she claimed to show how human nature was malleable rather than fixed. For example, she depicted young Samoan girls as sexually liberated, engaging in premarital sex (see Figure 9.1). This was shocking and intriguing to contemporary Europeans and Americans. The 1920s, remember, was an age in which women were supposed to remain virgins until they were married, and hook-ups were frowned upon. In 1935 she published *Sex and Temperament in Three Primitive Societies,* in which, among other things, she described the Tchambuli (more correctly Chambri) people, where the sex roles were reversed, with men decorating themselves and women being the practical ones catching and selling fish and using the money to provide for the family.

These apparently topsy-turvy cultures in which everything seems to be the other way around, led Mead to state "we are forced to conclude that human nature is almost unbelievably malleable, responding accurately and contrastingly to contrasting cultural conditions" (Mead, 1935, p. 280).

Figure 9.1 Three young Samoan girls photographed in 1902, 22 years before Mead arrived on the island.

Source: Ernst von Hesse-Wartegg, Publisher: Leipzig, J. J. Weber 1902, uploaded by Teine-savaii [Public domain], via Wikimedia Commons.

Mead's research has since been criticized. In 1983 – 3 years after Mead's death – Derek Freeman reported that the girls whom Mead interviewed in Samoa had admitted to making up stories in order to shock the young American anthropologist reverse (Freeman, 1983), while Frederick Errington and Deborah Gewertz (1985) found no evidence of reversed sex roles among the Chambri. Women did indeed do the fishing, but men still held sway in government and thus had the power.

Research (see next section) therefore seems to suggest that although human nature is certainly malleable, it is not as malleable as Mead suggested. For example, Joan Bamberger (1974), in *The Myth of Matriarchy*, argued that there is no evidence of a society in which men are subservient to women. Before moving on, we need to be clear that this should not be used as an argument to suggest that women should never be in power. This would be a variant of the **naturalistic fallacy** (see Chapter 1). There are doubtless many reasons why women have been historically subservient to men: men are generally physically stronger and more aggressive, for example. In postindustrial societies, which rely more on consensus and negotiation than dominance by brute force, the opposite could conceivably become the case, if it isn't already.

Are there any cultural universals?

A **cultural universal** is a practice or behavior that is common in all cultures studied. It does not mean that everyone adheres to it, nor that it is considered "good" by the people of that particular culture. There is a story of an argument between two Donalds: the cultural anthropologist Donald Brown and the evolutionary anthropologist Donald Symons. The argument was about the topic of this section – cultural universals. Symons was of the opinion that such things existed, whereas Brown, schooled in the tradition of Boas and Mead, believed that cultures are unique. Not wanting to duck the issue, Brown bet Symons that for any putative universal, he could find a culture in which the apparent universal was reversed or absent and set off to study the anthropological record to prove Symons wrong. He lost the bet, but at least got a book out of it. The fruits of his labor, *Human Universals,* detailed over 200 different cultural universals. A sample is shown in Table 9.1.

> **KEY TERM**
>
> **Cultural universal** A practice or behavior that has been shown to exist in every single culture studied (see Table 9.1 for some examples).

Table 9.1 Just some of Brown's cultural universals.

aesthetics	sexual jealousy
belief in supernatural/religion	sexual modesty
body adornment	incest, prevention or avoidance
childbirth customs	males more aggressive
childcare	marriage
copulation normally conducted in private	medicine
crying	melody
customary greetings	moral sentiments
dance	mourning
death rituals	murder proscribed
division of labor by sex	myths
fear of death	poetry/rhetoric
gossip	preference for own children and close kin
husband older than wife on average	snakes, wariness around
pretend play	socialization includes toilet training
pride	special speech for special occasions
rape (and rape proscribed)	statuses and roles
revenge	tabooed foods
recognition of individuals by face	tabooed utterances
right-handedness as population norm	territoriality
rites of passage	thumb sucking
rituals	tickling
sexual attraction	tools

Source: Data from Brown (1991).

To be clear, Brown's universals, such as those in the table, are "cultural" in the broad sense: there is no requirement that any of them are, in Hoebel's terms, "not the result of biological inheritance." In fact, some theorists, such as Steven Pinker, have proposed that some of these "cultural" practices are the result of some innately based human psychology. For example, in Chapter 4 we saw that the asymmetric costs of sex and childcare can lead to women tending to marry older men rather than the other way round, and, of course, the division of labor by sex. Childcare itself, as we saw in Chapters 1 and 6, is a result of humans' adoption of an extreme mammalian strategy in which future rather than current reproductive success is emphasized. The omnivore's dilemma, described at the beginning of Chapter 1, can lead to food taboos, and incest avoidance – found in many species – is a way of preventing children from getting two copies of harmful recessive genes, which is more likely when we mate with close relatives.

Some have less obvious evolutionary advantages and may be either a side effect of other physiological processes – thumb sucking is likely to be a hangover from the comfort-giving and evolutionarily vital infant activity of suckling – or neat ideas that spread from culture to culture – medicine, possibly. What is further interesting about this list is that the universals exist at a level of abstraction that makes them sometimes hard to notice. As humans we seem to be drawn to differences between people to the point where we fail to notice the similarities (maybe this is another universal?). So we notice that this person speaks Swahili, and that person speaks English; this person wears trousers, but that person wears robes; this person believes in many gods, but that person just one; this person sings using the diatonic (doh-ray-me) scale, whereas that person also sings the notes in between the "notes" (such as the Arabic quarter-tone system). What we invariably fail to notice – and what a hypothetical Martian might – is the surprising fact that all humans have language, body adornments, belief in the supernatural and music.

This is by no means to dismiss these differences as unimportant, they need to be understood, and they certainly are an important part of people's identity. It's just that sometimes we are so focused on the trees of difference that we fail to see the forest of similarity.

DISCUSS AND DEBATE – CULTURAL SIMILARITIES AND CULTURAL DIFFERENCES

Take some of the examples in Table 9.1 and try to list the different ways that different cultures do things (e.g., different food taboos). Might there be reasons for these differences? As a further challenge, consider whether the universals might be evolutionary adaptive, and why.

What is the point of culture?

Although culture is very probably not unique to humans (see Key concepts box 9.1) it is certainly something that, by and large, defines humans. So an important question is whether the ability to have culture itself actually evolved. Consider this: Most species are limited to the areas that they can inhabit as a result of their being adapted to local temperatures, foods or habitats. Humans, on the other hand, have managed to eke out a living in all but one of the continents on earth. The exception being Antarctica, which is not only very cold but also lacks drinking water for much of the year and, because of its isolation, would have been difficult for ancestral humans to populate.

Note that here we are discussing a species. Birds, for example, inhabit all continents including Antarctica (e.g., penguins), but birds are a *class* of animal, like mammals, not a species; all humans are member of the same species (see Chapter 2). The reason for humans' success is down to culture. Unlike polar bears, humans do not have thick fur to keep out the cold, so they simply take the thick fur from the polar bear and use that to keep warm. Using weapons such as spears and fishing rods, humans catch novel foods and use cooking, marinades and spice mixes to render them palatable. They build shelters to keep out of the bad weather and medicines to defend themselves against the local parasites. They keep detailed information of what foods are good to eat and what are bad, and create food taboos to protect against the nastiest or least obvious ones.

KEY CONCEPTS BOX 9.1

Do non-human animals have culture?

At first the idea of non-human animals having culture seems vaguely ridiculous. Baboon art? Monkey tennis? But, guided by Hoebel's definition of culture being practices that are not innate, researchers have considered the question more deeply. Being our closest relatives, the chimpanzees seem a good candidate and research has shown that chimpanzees do indeed seem to have something like culture. Whiten et al., (1999) have shown that there are many variations among different groups of chimpanzees that do not seem to be reducible to genetics. For example, some chimps use tools to pick termites from their mounds and others use rocks to crack nuts, and each group has different methods of doing this and some do not do it at all. There are also learned variations in grooming and courtship. Japanese macaques (a small primate) have learned to wash sweet potatoes in order to remove unpalatable grit (Kawai, 1965) and engage in snowball fights in the winter. Again, something they have apparently learned from one another. Outside of mammals many birds seem to acquire at least part of their song by imitating those around them (Catchpole & Slater, 2008), something which has led people to claim that such birds have regional dialects (Catchpole & Slater, 2008).

The more we look the more it looks like culture is not unique to humans although the examples are much more basic than human examples of culture. Monkey tennis is still a long way off.

So what is culture for? Robert Boyd and Peter Richerson (Boyd & Richerson, 1983; Richerson & Boyd, 2001; Boyd, Richerson & Henrich, 2011) have an idea. They suggest that culture – or rather the ability to acquire culture – evolved by a process of **natural selection** in order to give humans a helping hand in coping with rapid environmental change. For the past 2.6 million years the earth has been in the grip of an ice age. It still is, as there is permanent ice around the polar regions. Prior to this there was no permanent ice, and prehistoric animals were able to inhabit the then-temperate polar regions. Throughout the past 2.6 million years there have been periods of comparative warmth, as now, punctuated by periods where ice sheets have covered the northern parts of North America and Northern Europe. Many other animals adapted to these conditions through a process of natural selection, for example, by evolving fur coats and other adaptations (see Figure 9.2), but, Boyd and Richerson suggest, our ancestors may have evolved the ability to create culture, including the ability for intellectual innovation and the ability to

Figure 9.2 Many animals adapted to ice age life by developing thick fur coats, such as the wooly rhinoceros.

Source: Image copyright © Ozja/Shutterstock.com.

learn these innovations from one another rapidly. This theory is called **dual inheritance theory** because they propose that in addition to the standard means of inheritance – through **genes** – humans have acquired a second means of inheritance, which is to acquire culture from our elders.

The mechanisms that underlie culture

There are broadly two kinds of mechanism that underlie the process of cultural transmission: cognitive ones, which permit it to happen, and motivational ones, which ensure it does happen.

Cognitive mechanisms of cultural transmission

One of the most obvious mechanisms for culture is the ability to imitate. Although frequently condemned – we talk about learning things "parrot fashion" or being a "copy cat" – imitation really is cognitively complex. A famous example is of psychologists Winthrop and Luella Kellogg (Kellogg & Kellogg, 1933), who tried the unusual experiment of rearing a baby chimpanzee, named Gua, alongside their own child, Donald. Their goal was to see if, given the correct environment, Gua would imitate Donald and become more human-like, even, perhaps, learn to speak. It didn't work out like this; instead of Gua copying Donald, Donald aped the ape: running about on all fours and imitating the "pant-hoot" signal that chimpanzees give when excited.

Humans are simply better imitators than chimps. Research by Nagell, Olguin and Tomasello (1993) reveals that adult chimps have the imitative ability of a 2-year-old human. At least one reason why humans have this advantage is that humans are capable of not just representing the *actions* of another, but can also *represent* their *intentions*. Or, to put it another way, in addition to being able to determine *what* someone is doing, they can also determine *why* they are doing it. So, if you are watching someone assemble a collection of sticks in order to reach some fruit and the person drops one of the pieces, you will probably realize that this is a mistake and omit the dropping when it comes to your turn. Chimps include the dropping stage, suggesting that they find it difficult to decide which actions matter and which do not (Tomasello, 1999). Representing intention means that humans are able to learn more efficiently by ignoring irrelevant steps, and it also means that they might be able to find quicker or more effective ways of achieving the same ends, aiding creativity. This ability to represent intentions is known as **theory of mind** or **mindreading** (Baron-Cohen, 1995).

Motivational mechanisms of cultural transmission

There is a wealth of research in social psychology showing how people adopt the practices, values or otherwise comply with or obey other people. Stanley Milgram's (1963) famous "obedience studies" of the 1960s show that people will comply with authoritative commands to give electric shocks to a learner as punishment for getting questions incorrect, even when it leads to them apparently killing a participant in the study (the "learner" was a confederate). Solomon Asch's (1951) conformity studies show how people will apparently disbelieve the evidence of their own eyes by falling into line and saying that a shorter line is more similar to a target line than one that is exactly the same (see Figure 9.3). And Philip Zimbardo's Stanford prison study revealed how participants will adopt the roles of prison guards and act abusively toward other participants who are playing prisoners (Haney, Banks & Zimbardo, 1973).

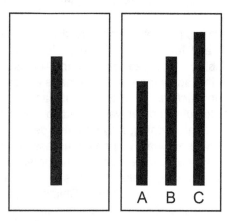

Figure 9.3 Asch's lines.
Source: Adapted from Asch (1951).

What these and many other studies show is that people not only have the ability to imitate other people but they are also strongly motivated to conform to their ways of doing things. Such behavior can be observed on the Internet, where people tend to choose to watch items that are listed as popular, making them yet more popular and thus leading to more people choosing them (Salganik, Dodds & Watts, 2006).

People conform even when they perceive themselves as being nonconformists – such as having a tattoo – where they imagine they are being a rebel but are actually conforming to a current social norm. In fact, if you want to look like everybody else today, the best way is get a tattoo.

A recently discovered phenomenon known as **pluralistic ignorance** provides a further example of people's desire to fall in line with others. *Pluralistic ignorance* is the finding that people often act the way they do, not because they genuinely believe it is right but because they believe that everyone else believes it is the right thing to do. This can result in the rather bizarre situation where a group of people act in a particular way, even though no one actually wants to do it. An everyday example of this is when a couple are deciding on where to eat. He would like to go for a pizza but believes that she would prefer to eat Thai. She, on the other hand, would also like to eat pizza but also believes that he would prefer Thai. During the subsequent discussion, she thinks that his insisting that what he really wants is pizza is just him being nice to her and taking her interests into account. He thinks the same thing when she says that she also wants pizza. They eventually eat Thai, each party feeling that they are doing what their partner wanted, but in fact, neither is.

Experimentally, this has been shown in a wine-tasting study in which participants were given three "fine wines" to sample (actually, the same wine with one spiked with vinegar). An "expert" (actually a confederate) pronounced the vinegary wine the best, when asked to state their preferences publicly the participants agreed with the expert, but when asked to state their opinions privately, they admitted they hated it (Willer, Kuwabara & Macy, 2009). Like the "Emperor's New Clothes," people will often act in accordance with opinions that they do not hold in order to avoid looking foolish, uncool or incurring the wrath of their peers.

The motivation to act in accordance with the view (or perceived view) of those around us is extremely strong and is certainly an important mechanism whereby cultural norms become established within a group. However, it is not clear whether these motivations evolved as part of a "cultural acquisition device" or whether they originally evolved for some other purpose such as maintaining cohesion in groups.

KEY CONCEPTS BOX 9.2

The scars of culture: Has culture changed our biology?

In this chapter we have argued that culture is rooted in biology in that the mechanisms that permit it have evolved. But a rather different question is whether culture has changed us biologically in any way. The answer is yes, but we are still working out the extent to which it has done so. One clear example is that of lactose tolerance. You might know people who are lactose intolerant (you might even be lactose intolerant yourself). If you don't know what this is, some people have an inability to process a sugar (lactose) found in milk, meaning that if they consume milk or some of its products they become sick. Although in the West lactose intolerance is relatively rare, across the world it is much more common, and if we consider the rest of the mammals it is lactose tolerance that is rare, not intolerance. Milk is a food for babies and, once weaned, most mammals lose the ability to digest it – essentially the gene for producing the enzyme that digests lactose – lactase – switches off. Somehow, *some* humans seem to have avoided this big switch off and can digest lactose in adulthood. Why? A clue can be found by looking at the ethnic origins of the people who can and can't consume milk. Most of the people who can digest milk have ancestors who kept cows, sheep and goats.

Essentially, these animals are ways of turning indigestible (to humans) grass into food, in the form of meat. But although you can only kill a cow once, you can use it for all its life as a grass-processing factory turning grass into another food, milk. For such people, milk and its products (cheese, butter, yogurt) became a vital source of food. This placed a selection pressure on humans, those with the relevant mutations for processing lactose lived and passed on the gene for this, those that could not process it were less likely to survive and lactose tolerance flourished (Bloom & Sherman, 2005). This kind of change in known as *gene–culture co-evolution*.

Primatologist Richard Wrangham (2009) has further argued that the cultural practice of cooking – which may have been discovered by one of our ancestors, *Homo erectus* (see Chapter 2) – has also left its mark on our genome. According to Wrangham, cooking food facilitates digestion, which means that *Homo erectus* needed to spend less time chewing and digesting food, freeing up time to do other things (humans spend much less of their day eating than their close relatives the chimpanzees do).

Cooking also means that *H. erectus* could evolve a much shorter gut than its ancestors, and this more energy-efficient gut freed up energy to run a larger brain, with all the implications that large brain size has for human behavior and culture.

From culture to civilization

So, the story so far is that certain cognitive and motivational mechanisms evolved (or were **coopted**) by natural selection in order to facilitate imitation.

Language and culture

We mustn't overlook the importance of language in the development of culture. Arguably, chimps and Japanese macaques can imitate in a "monkey-see, monkey-do" fashion, but what they cannot do is give and follow instructions. With language, someone can tell you what to do without showing you; they can also explain specifically *why* they are doing something in a particular way. Without language, it is near impossible to pass on information, such as "Don't eat those berries, they are poisonous" or "If you are ever bitten by a snake, perform the following procedure. . . ." Language is also important in the coordination of action, for example, in group hunting.

Are humans a superorganism?

Taken together, humans' superior imitation abilities and ability to use language meant that individual discoveries could be quickly owned by the group. This means that humans are so much more than a collection of individuals. To borrow a term from the Boas school of anthropology, it allows humans to become a kind of superorganism. Let us briefly consider what this means. The majority of the cells in your body have given up reproduction and, instead have delegated this responsibility to your sperm or egg cells. The net result of this is to ensure cellular cooperation. Any body cells that evolved a tendency to go *off-piste* could potentially affect the ability of the body to survive, which would affect their own reproduction. Considered this way, an ants' nest can also be considered a single organism made up of closely related elements (the individual ants), all of whom can only reproduce indirectly through the queen. For the same reason body cells tend not to go AWOL, ants don't either. And like body cells, ants have evolved specialisms. Just as there are specialized liver cells, skin cells and nerve cells, so there are soldier ants, nurse ants and worker ants.

Humans are different. Each individual has the capacity to reproduce on its own; humans have not delegated reproduction to a queen (not even in England), so, cooperation in humans is not guaranteed (see Chapter 5). But, because of the mechanisms discussed above that underpin culture, we have specialized individuals: humans can be doctors, lawyers, singers, drivers and football players – we are no longer "generalists" like our ancestors or most other mammals. To be clear, we have not evolved into specialists like ants, but culture and the psychological mechanisms that underpin it have enabled this situation to develop over thousands of years. And, it is another reason for our success (where success is simply defined by our sheer numbers and distribution across the globe).

The economics of specialization

The British evolutionist, Matt Ridley, poses the following problem.

> Imagine that there are two people Adam and Oz. Adam takes four hours to make a spear and three hours to make an ax (total time 7 hours). Oz, on the other hand, takes one hour to make a spear and two hours to make an ax (total time 3 hours). Oz is better at both spears and axes so does Oz need Adam?
>
> (Ridley, 2010)

Before reading on, try to answer the question.

On the face of it, it seems that Oz does not need Adam as Adam is worse at both spears and axes than Oz. But, consider further – although Oz is better than Adam at both, he is MUCH better at axes. So, think what would happen if Oz could focus exclusively on spears and persuade Adam to focus on what he is best at (but still worse than Oz): axes. Oz could then make two spears in 2 hours (a saving of 1 hour), whereas Adam could make two axes in 6 hours (again, saving an hour). Once made they then trade, so each has an ax and each has a spear.

This is the logic of specialization: it is better to delegate tasks that you are less good at to focus on tasks that you are better at, even if the person to whom you are delegating the task is worse than you! And, of course, because you are specializing in one thing, you will probably get even more efficient at producing it (Ridley, 2010).

Specialization and a trade

Central to specialization is, of course, trade. One cannot live on axes alone, and promises made through language are one way in which items not needed can be exchanged for items that are. It has even been suggested that one of the reasons why anatomically modern humans flourished at the expense of Neanderthals is because humans traded. There is no evidence for Neanderthals engaging in trade, but there is plenty for contemporary humans (Gamble, 1999; Horan, Bulte & Shogren, 2005).

Fast-forward several thousand years to the present, and specialization has reached such an extent that almost literally no one knows how to make anything anymore. A point first made in a whimsical essay by the economist Leonard Read (1958). This essay is written from the perspective of a pencil who tells us that no one single person knows how to manufacture his kind. Think of it. Trees need to be grown, chopped down and manufactured into the body. Graphite needs to be mined and mixed with clay, which also needs to be mined, to make the lead. The "rubber" of the eraser needs to be harvest from plants, vulcanized and mixed with pumice (which is what does the erasing, the rubber merely holds it all together) and so on and so forth.

A further point of the pencil's story is that all of the above happens in many countries to the extent that even the humble pencil is a genuinely global product: no industrialized country in the world is truly self-sufficient. This sometimes worries people, but as American evolutionist Robert Wright (2001, 2007) joked there are many reasons for not declaring war on Japan, and one of them is that they made his car. His point is that interdependencies between nations are potentially a source of peace rather than a cause for concern.

DISCUSS AND DEBATE – DEPENDENCIES AND INTERDEPENDENCIES

We often consider that self-sufficiency is a good thing, but as Robert Wright points out, being dependent upon other nations could actually be a good thing for global peace. Do you think he is correct in this regard?

KEY CONCEPTS BOX 9.3

International differences in wealth

People often ask why some countries – traditionally many of those in Africa and South and Central America are relatively poor, and until recently technologically less advanced than Europe, the United States and parts of Asia. Although there are doubtless many political reasons for this, Jared Diamond (1997) attempts to answer the question by going way back in history to the beginnings of civilization. The beginnings of civilization lie in agriculture, which creates specialist food producers, which free up time for others to get on with making weapons, farming implements and writing books of philosophy. Oz (see above) can only specialize in making spears if he doesn't have to spend hours hunting or gathering food but can simply exchange some for a spear. Agriculture began in the area known as the *fertile crescent*, near where modern Iraq is now. But why there? Diamond shows how Eurasia (the combined continents of Europe and Asia – they are really only one landmass) has the correct species of plants and animals for domestication. Food animals such as sheep, cows, pigs and chickens are all Eurasian animals, as are wheat, beans, peas and many other crops. Horses (as well as oxen) are also Eurasian and could be used to pull plows and other machinery in order to more efficiently harvest and plant crops. By contrast, there were no domesticable animals in Africa, North America and only the Llama and related animals in South America.

So, the existence of the correct species for domestication led to agriculture, which led to specialization, which led to ships, steel swords, infantry, cavalry, a navy, guns and a desire to own the rest of the world in order to support its burgeoning population. Diamond's book is named after the tools that enabled this colonization process: Guns, Germs and Steel. The significance of the first and last have already been dealt with, but what about germs? This is the result of domesticating animals. Many of the deadliest diseases in history have jumped the species barrier. Chicken pox, smallpox and influenza all came from domesticated animals. By accident, farming also produced one of the deadliest weapons of Eurasia: disease. When Eurasians invaded other counties, they often killed as many people through the diseases that they brought as with guns and swords. The people they invaded had no contact with domestic animals, and hence no natural immunity. If evidence were needed for the destructive nature of disease, the 1928 "Spanish" influenza pandemic killed 50–100 million people (3%–5% of the world's population). Even now, many of the most threatening diseases, such as forms of avian influenza, come from pathogens crossing the species barrier in agricultural communities – often in China, where the inappropriately named Spanish flu epidemic originated.

The future of cultural studies

Mainstream psychology has generally not taken much interest in culture. While recognizing it as a very important influence on development (see Chapter 6), it has tended to leave the study of culture to anthropologists and sociologists. This really does need to change, and we hope that this chapter has helped you to see why. Most people who research these things agree that our brains have changed little in the past 10 or so thousand years, but in that relatively short time our species has gone from hunter–gatherers living in relatively small communities of possibly 150 individuals (see Chapter 2) to latte-drinking sophisticates who have little idea what our food is, let alone where it came from and who killed it. These differences might be on the surface, but what a surface.

Summary

- Culture has traditionally been thought of as a force that is separate from biology, but recent evolutionary thinking has asked the question: What is culture for? One answer is that it enables humans to adapt much more rapidly than by biological evolution in order to respond to changing environmental conditions. This is known as *dual inheritance theory*.
- Early attempts to study cultural differences attributed them to biological differences. Things changed when Franz Boas established cultural relativism as a method in social anthropology. This aimed to understand cultural practices within the context of the wider culture. The Boas–Mead view that humans are malleable and cultures infinitely variable has been shown to be incorrect. Brown's work on cultural universals shows a high degree of commonality across different cultures, even though they may differ in detail.
- Mechanisms such as the ability to imitate are essential for ideas to spread through a culture, and humans, it seems are extremely good imitators. Humans' tendency to conform and obey authority is also an important motivator for the adoption of cultural practices. There is debate as to whether imitation evolved in order to permit culture directly, or whether this ability was coopted.
- Culture as left its mark on the genome. Examples are lactose and cooking. There are likely to be many more, once we start looking properly.

- Cultural specialization and trade have been important in the development of large-scale civilizations, leading to humans' domination of the planet. One hypothesis argues that our direct ancestor's ability to trade may explain why they out-competed Neanderthals.

FURTHER READING

Richerson, P.J., & Boyd, R. (2008) *Not by genes alone: How culture transformed human evolution*. Chicago: University of Chicago Press. Readable book explaining Boyd and Richerson's dual inheritance theory.

Diamond, J. (1998) *Guns, germs and steel: A short history of everybody for the last 13,000 years*. London: Vintage. A really interesting read. Attempts to explain differences in cultural wealth across the planet.

Darwinian differences

10

What this chapter will teach you

- The history of personality measurement.
- The question of personality in animals.
- The question of the involvement of genes in personality.
- The relationship between evolution and intelligence.

Why do we differ?

It has been said that if a Martian were to visit the earth, humans would appear to be pretty much the same as one another. There would, of course, be some observable differences, such as sex differences, slight differences in coloration, clothing and, of course, language, but in the main these differences would be vastly outweighed by the similarities. In fact, if we are speaking about genetic diversity, then recent evidence suggests that humans are substantially *less* diverse than their close relatives, the great apes, including the chimpanzees (Kaessmann, Wiebe, Weiss & Pääbo 2001). This is surprising. Genetic diversity within a species usually increases over time as there are more opportunities for mutations to occur, given that our species' lineage is the same age as both chimpanzee species (around 5–6 million years; see Chapter 1), humans and chimps should be approximately similarly diverse. One theory for the low diversity of humans is that a "bottle-neck" occurred comparatively recently (30,000–130,000 years ago), and the human population dwindled to a small number of individuals. This had the effect of resetting the clock, and diversity has had only a comparatively short time to recover (see Premo & Hublin, 2009, for a discussion of relevant theories).

It certainly doesn't seem like this to humans. We obsess on our differences, and psychologists are no different. Why are some people shy? Why are some people prone to depression? What makes a genius? A psychopath? A martyr? As evolutionists we might ask a more general and more fundamental question, what we have referred to as an "ulti-mate" question, and it is this: Is there an evolutionary benefit for people to be different from one another, and if there is, what is this benefit?

DISCUSS AND DEBATE

Before you read on, think about (or discuss in a group) the potential benefits for humans being different in the ways that psychologists normally discuss – for example, extraversion, optimism or intelligence. Remember, benefits in evolutionary terms might be cast in terms of survival, reproduction and childcare.

In discussing the benefits of survival with students, one of the most common answers given is "that it wouldn't do for us all to be the same." While this is not necessarily an incorrect answer (see later in this chapter), it is rather vague. In what ways might it be disadvantageous in terms of survival and reproduction for us all to be the same in terms of personality and intelligence (for example)? How might, for example, a shy person do better if they are with a certain number of extraverts than if everyone was shy (and vice versa)?

Before we go on to describe the various accounts for the existence of individual differences, it is worth exploring their more immediate causes so that we know what we are discussing.

In Chapter 6 we discussed two broad sources of individual differences: those caused by **genes** and those caused by the environment. We further suggested that differences in personality and intelligence seem to be between .3 and .7 **heritable,** meaning that somewhere between 30% and 70% of the **variation** among people is the result of genes. One of the challenges of an evolutionary account of individual differ-ences is why some of the variation seems to be heritable and some of it is not. This is a question that we will return to later in the chapter.

The measurement of personality: A little history

The physical world contains a lot of different kinds of matter. There is wood, iron, air, stone, flesh, water, bone, sand and many, many other kinds of material. Early physicists wanted to find out whether these tens of thousands of materials could be reduced to a small number of simpler entities. The first big step was to reduce all matter to what is now around 100 elements (they are called *elements* because they are simple entities). Further research showed that elements themselves could be reduced to simpler entities: protons, electrons and neutrons, which themselves could be reduced to a larger family, including quarks, gluons and other subatomic exotica.

A similar approach was taken to personality by the American psychologist Gordon Allport in the 1930s (Allport & Odbert, 1936). He observed that there were a vast number of words used to describe personality; for example, the English dictionary contained 17,953 different words, such as *outgoing, optimistic, shy* and *anxious.* His task was to attempt to find, if you like, the fundamental elements of personality that acted as the building blocks for all these different **traits**.

First, he observed that many of these 17,953 words were either synonyms or close synonyms of one another (*shy* vs. *reserved,* for example), so, by grouping together synonyms it is possible to reduce this rather large number to a smaller, more manageable number. Furthermore, some of the words are antonyms, such as *shy* versus *gregarious.* You cannot, of course, simply treat antonyms like synonyms and group them together because they are opposites, but what you can do – and what Allport did – was to create a series of personality dimensions (see Figure 10.1)

We can see in our simple example that we have reduced 4 personality words to one single personality dimension, which – following tradition – we might call "extraversion" (people judged high on the right-hand words would be called *extraverts,* those judged high on the left-hand side would be called *introverts*).

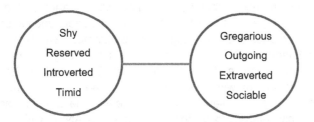

Figure 10.1 An example of a personality dimension. Synonyms are grouped together (inside the circles) and are placed in opposition to antonyms.

Statistical ways of creating personality types

Although the above process can certainly reduce the vast number of personality words to a smaller number, it is rather subjective as to what constitutes a synonym or antonym so more recent attempts try to squash the large number of potential **personality traits** into a handful of different personality types by using statistical means. The method – known as *factor analysis* – is mathematically quite complex and beyond the scope of this book to explain in detail. In essence, we devise questions that are thought to measure particular personality traits, such as shyness and neuroticism, and hand these out to a large number of participants for completion. These data are then fed into a computer, and we look for correlations among questions. Questions that are either negatively or positively correlated are grouped together to form a single personality dimension. This is essentially what we did above, but in a less subjective way. Questions that are not correlated with one another are thought to belong to different personality dimensions. The end of this process might result in a few tens of different dimensions. The next step is to take these recently formed dimensions and correlate *them* with each other. Correlated dimensions are again collapsed together and the process repeats. At the end of this process we will have a few personality **factors** that are thought to be fundamental to human personality. In most measures of personality, each factor is unrelated (not correlated with) any of the others factors. For example, how shy you are is unrelated to how anxious you are. Why? Because if shyness were related to anxiety, then this could be collapsed to form a new factor that encompasses both shyness and anxiety; remember the purpose of the exercise is to find the fundamental "elements" of personality, in the same way that chemists did with the periodic table.

How many factors?

In truth the above process of creating personality factors by lumping together dimensions that are correlated with one another does involve a certain degree of subjectivity about which factors one keeps distinct and which ones one lumps together. For example, British psychologist Hans Eysenck (1990, 1991) found three personality factors: extraversion, neuroticism and psychoticism, but Costa and McCrae (1985, 1990) ended up with five: extraversion, neuroticism, openness to experience, agreeableness and conscientiousness (these factors are commonly referred to by the acronyms OCEAN, CANOE or, rather less helpfully, NEOAC). These differences in factor number aren't arbitrary or accidental, the result of too much or too little cranking of the factor-analytic handle, but reflect differences in the aims of the researchers that produced them. To adherents of the Costa and McCrae's Big Five model, it makes more sense to partition Eysenck's extraversion into extraversion and openness, and Eysenck's psychoticism into conscientiousness and agreeableness as they are better able to discriminate more subtle forms of behavior (in this case someone low on agreeableness and low on conscientiousness would come out high on psychoticism). Evolutionist David Buss has suggested that including curiosity (openness) in the Big Five it makes the scheme more evolutionarily plausible as curiosity is an important trait in many animals (including non-human ones, as we shall see).

In truth, then, no current set of personality factors represents anything like the fundamental elements of personality, analogous to the theory embodied in the periodic table. Whether one has three, five or more (or fewer) factors is more to do with the usefulness in predicting behavior. To this end, many scales contain a number of subscales. For example, for the Big Five the 60 item NEO PI-R (Costa & McCrae, 1985) lists six subscales for each of the five factors (Extraversion, for example, is made of up the following subscales: Warmth, Gregariousness, Assertiveness, Activity, Excitement Seeking and Positive Emotion).

KEY CONCEPTS BOX 10.1

Sample questions from the Big Five personality questionnaire

The following are a sample of the questions relating to each of the dimensions in the Big Five personality questionnaire.

Openness to experience

> I spend time reflecting on things.
> I am full of ideas.
> I am not interested in abstractions. (*reversed*)
> I do not have a good imagination. (*reversed*)

Conscientiousness

> I follow a schedule.
> I am exacting in my work.
> I leave my belongings around. (*reversed*)
> I make a mess of things. (*reversed*)

Extraversion

> I start conversations.
> I talk to a lot of different people at parties.
> I don't talk a lot. (*reversed*)
> I keep in the background. (*reversed*)

Agreeableness

> I feel others' emotions.
> I make people feel at ease.
> I am not really interested in others. (*reversed*)
> I insult people. (*reversed*)

Neuroticism

> I worry about things.
> I am much more anxious than most people.
> I am relaxed most of the time. (*reversed*)
> I seldom feel blue. (*reversed*)

The stability of personality factors over different situations

Personality traits are often contrasted with **personality states**. The former refer to relatively enduring ways of behaving, whereas the latter refer to ephemeral and situation-dependent ways of behaving; for example, we all feel anxious from time to time, depending on context, such as before an exam. In order for us to be certain that personality traits exist, we need to show that people are relatively stable in the way that they behave from one situation to another (i.e., that we are measuring traits rather than just states).

One of the first psychologists to question this assumption was Walter Mischel. After reviewing the evidence, in 1968 he found that when behavior relevant to personality was investigated (e.g., outgoingness, honesty) people behaved somewhat differently across different situations. In other words, a person who acts honestly in one situation might act dishonestly in a different situation. The argument as to whether traits are really stable has become known as the **person–situation** debate and is concerned with whether the primary determiner of behavior is a person's personality or the specifics of the situation. The debate has continued over the years (see Mischel, 1968; Ross & Nisbett, 1991), but it is important that it is properly understood: no one doubts that there is *some* level of consistency across situations – there are always some effects of personality – the key question is to what extent personality matters in guiding behavior.

From an evolutionary point of view, the fact that humans are flexible in the way that they behave should not be too surprising. As with other animals, it is important that humans are responsive to the particular situation that confronts them. To this end anxiety, honesty and sociability should depend upon the individuals goals. For example, if threatened, be anxious; if being dishonest can get you out of a hole, consider that as a tactic; if you are trying to impress someone, then being sociable might be a good way to behave.

Personality in non-human animals

Some people balk at the very thought of animals having a personality. Perhaps because it makes them think of those people who swear that their pet cat understands what they say or can read their mind. Many psychologists are also wary of using personality to apply to non-humans. After all, one of the things that all psychology students are warned about is the dangers of anthropomorphizing – imbuing non-human animals with human characteristics (see Chapter 7). This is, of course, entirely appropriate, and it is entirely inappropriate to assume that cockroaches have compassion or that dogs have language just because humans have such capacities. On the other hand, one can go too far. If a non-human behaves as if it has compassion, if it helps out non-relatives who are suffering, for example, if it shows signs of distress when confronted with suffering – in short, if it does all the things that humans do when we say that they are showing compassion, then why not use the word? Put another way, anthropomorphism is inappropriate when one simply assumes *without any evidence* that non-human animals have similar qualities to humans, but if there *is* good evidence, then surely to refuse to believe that animals have that quality is just a form of human prejudice.

So, is there any evidence that non-human animals have differences in personality? In fact, there is. Research by O'Steen, Cullum & Bennett (2002), has shown that guppies (a small tropical fish that are often kept as pets) that live in fast-flowing water are bolder than those living in calmer environments. We cannot, of course, measure boldness by administering a personality questionnaire such as the Big Five, but it can be measured behaviorally. Bolder guppies are more likely to explore an unfamiliar environment and approach a potential predator than those who are less bold.

The researchers further suggest that these differences are not merely arbitrary but have a purpose. Fast-flowing water is less likely to contain predators such as pike. In the absence of such predators, it pays guppies to be bold as this provides them with opportunities for obtaining food and for mating. In slower flowing waters, the increased likelihood of predation can be a disadvantage: a bold guppy is more likely to become dinner.

Of course, it could simply be that guppies in predator-rich waters have learned to curtail their boldness, but this seems not to be the case, or at least not the full story. When the researchers placed the offspring of guppies from populations in fast-flowing waters into pike-infested territory, they show the same boldness and as a result are much more likely to be eaten. Like human personality, boldness in guppies seems to have a genetic component, and timidity seems to be an antipredator adaptation.

Similar findings were obtained in research on the great tit (*Parus major*). Like guppies, great tits vary in their degree of boldness: high scorers are bold, inquisitive and show high aggression; low scorers are more reserved (Dingemanse, Both, Drent, Van Oers & Van Noordwijk, 2002). This trait is also heritable, meaning that parents pass this trait on to their offspring (estimates of heritability range from .3 to .6 – see Chapter 2).

Unlike guppies, however, the evolutionary benefits of boldness and timidity are different depending on whether the bird is a male or a female. For females, being bold benefits them when food supplies are scarce as such individuals are more likely to search and find food supplies. When food is plentiful, however, the extra risk of predation of moving farther afield and reduced interspecies conflict tends to favor those who are more timid.

Males show the opposite pattern. As for many birds, great tits will often fight with other males to compete for females. When there is little food, death from starvation means that there are comparatively few other males to compete with so timid males can gain access to females without engaging in potentially damaging conflicts with other males, which is what the bolder males will do. When food is plentiful there are more males and so access to females requires competition. In such a case it is the bolder ones that, on average, tend to do better (Dingemanse, Both, Drent & Tinbergen, 2004).

DISCUSS AND DEBATE – ANIMAL PERSONALITY

Do you think it makes sense to think of animals as having a personality? Can you draw on any of your own experiences to support or contradict this idea?

Is there an evolutionary benefit to being different?

As we discussed in Chapter 6, most of the frequently used personality factors (extraversion, introversion, neuroticism) seem to be heritable, meaning that around 50% of the variation among people is explained by genetic differences. The fact that these individual differences are partially genetic in origin has led evolutionists to speculate that there may be some evolutionary advantage to personality differences. Here we discuss what they might be.

If we consider human rather than non-human animals, then there are a number of possibilities as to whether or not individual differences in personality are adaptive in terms of advantages in survival and/or reproduction.

No advantage

The first possible explanation is that differences in personality have no adaptive consequences. As we know, sexual reproduction involves the recombination of half the genes from each parent. You might think of this as a game of cards. Two people (the parents in this analogy) each have a hand of cards. As part of the game, each person has to give a third person (the offspring) a randomly chosen sample of their cards, amounting to half of their cards. Sometimes the third person gets a great set of cards – full of flushes and runs – sometimes an OK set of cards and sometimes a terrible set of cards. A good set of cards here represents a set of genes that work well together whereas producing a poor set of cards represents genes that work less well. Some people's personalities might indeed be suboptimal, leading to negative fitness consequences, but remember it is the combination of genes or cards that matters, not the individual cards/genes themselves. Depending on the card game, a two of clubs can be an excellent card to get (e.g., if you have an ace of clubs, a three of clubs and a four of clubs) or a lousy card (if you have three queens and desperately want a fourth). Because the fitness of a gene, is dependent upon other genes that were inherited there would be no selective pressure to remove it from the gene pool. We saw a similar argument in Chapter 8, relating to mental illness and also sickle cell anemia genes which are kept in the population because they have beneficial effects when paired with a normal gene (immunity to malaria) but negative effects when in the presence of another sickle cell gene (the sickle cell disease). Although the gene negatively affects some people it remains in the population because of the positive effects in others. Returning to personality, some people be disadvantaged by inherited personality traits (e.g. by being cripplingly shy) but the genes may only cause shyness when paired up with certain genes, when with other genes they may give fitness benefits.

Changing environmental hypothesis

This hypothesis, used to explain variation in human personality, is perhaps closest to the research on guppies and great tits. Recall that bolder female great tits are favored when food is poor (because they are more likely to obtain food) and disfavored when food is plentiful (because they are more likely to be preyed upon). If food was always poor, then we might expect timid females to gradually be weeded out by natural selection; if food was always plentiful, then boldness might become the disfavored trait. However, if the environment varies in the availability of food, then in some years bolder animals are favored and at other times more timid animals are favored, meaning that the population will always be mixed.

It is possible that this might account for some of the individual differences in human personality. In Chapter 9 we discussed that fluctuations in climate, such as during ice ages, may explain the evolution of culture (see Boyd & Richerson, 1983 – Chapter 9) they may also account for particular kinds of personality types. Ice ages may have favored personalities that were more sociable and community-minded – one for all and all for one – in order for each individual to survive. However, ice sheets didn't cover the entire surface of the globe (see Figure 10.2), so we would have to explain the existence of individual

Figure 10.2 Map depicting the extent of the ice sheets during all recent ice ages.

Source: Image created by Wm. Robert Johnston [Public domain], via Johnston's Archive (http://www.johnstonsarchive.net/spaceart/cylmaps.html).

differences in those areas where the ice ages did not penetrate. Potentially, we didn't need ice ages to lead to changes in food supplies, as for the great tits. Maybe there is enough variation outside of ice-ages in order for different personality types to be favored.

KEY CONCEPTS BOX 10.2

Do genes code for personality?

The nature–nurture debate is an enduring theme that runs throughout psychology. Today it is well accepted that our behavior and internal states are the result of a complex interaction between genes and environment. As we saw with notions of heritability, however, there is still a robust debate regarding the relative effects of genes and environment on human behavior. In recent years, due to technological advances, a new field of study – **behavioral genetics** has been developed in order to help answer this question. Over the last 15 years, new discoveries of genes for various aspects of personality and intelligence have hit the headlines. These include claims of genes for criminal behavior, intelligence, depression and warmth (Kogan et al., 2011). But, what does it mean to suggest that there are genes for particular features of personality? And how well does this notion stand up to scrutiny?

The first thing we have to realize is that to state there is a gene for a particular trait is a gross simplification (see Chapters 2, 6 and 11). What behavioral geneticists actually study is the degree to which differences between people can be attributed to difference between them in their genes. Second, we have to accept that genes that code for **phenotypic** effects always do so within a given environment – change the environment and you may well alter the effects of the genes. Third, although claims have been made for single gene effects, genes often interact in complex ways, which means other genes may enhance or counteract the effects of these genes.

Having established these provisos, what can we say about the state of play with regard to the relationship between genes and personality? A number of single genes have been identified that occur in different versions (**alleles**) that appear to play a role in differences in personality between people. Such genes generally have effects on how well **neurons** (nerve cells) pick up **neurotransmitters** (chemicals that cause neurons to alter their firing rate) and, hence, on behavior. These are known as **candidate genes**. Two such candidate genes are labeled D4DR and 5-HTT. The D4DR exists in various forms and is believed to affect how well the neurotransmitter dopamine is picked up by certain specific brain cells and, hence, how well these neurons utilize dopamine (Eichhammer et al., 2005). One particular form of the D4DR is considered by some experts to lead to an increase in sensation-seeking (extravert) behavior. In the case of 5-HTT, one form of the gene is claimed to be associated with anxiety-related problems such as shyness, obsessive-compulsive disorder and bulimia nervosa (an eating disorder). Again, the effect is believed to be related to how well a particular neurotransmitter is utilized in the brain (Monteleone et al., 2006).

In addition to these genes that have been claimed to affect levels of sensation-seeking and anxiety, one candidate gene (that has recently caused a bit of a stir) has been labeled by some the "empathy gene." This "empathy gene" exists in different versions, and people who have the genetic code of GG at a particular locus are seen as more trustworthy, kind and compassionate when compared with those who have AG at this locus (Kogan et al., 2011 – see Chapter 2). The researchers were able to establish this by looking at judgements made by other people when observing GG and AG individuals during social interactions. It is believed that the increase in what we can broadly call "observed warmth" occurs in GG people because this version of the gene affects how well **oxytocin** is utilized in the brain. The discovery of this gene made headlines in the international media around the world – but, bearing in mind the above provisos, we really need to be careful not to jump to the conclusion that kindness is all about having this version of a particular gene. The researchers were careful not to label their find the "empathy gene" – that came from the press. In fact as with other genes that have been implicated in differences between people on personality measures they were keen to stress that this gene has an effect that relies on interactions with other genes and environmental factors, such as life-experiences. This is also true of the D4DR and 5-HTT genes – the effects of both of which have been disputed (Flint, Greenspan & Kendler, 2010). Hence, the degree to which specific single genes influence aspects of personality, from anxiety to empathy, is an area for which behavioral geneticists still have their work cut out.

Trade-offs

Perhaps a more plausible account for individual differences relates to the fact that there are multiple strategies for ensuring that genes survive into future generations. We discussed one strategy, the **C-F continuum,** choice in Chapter 6. Remember that this relates to the extent to which an individual opts to maximize their current fitness – having a lot of children but investing comparatively little in them – or future fitness – having fewer children but investing in them more – (Chisholm, 1999; Belsky et al., 2010). Each of these strategies represents a method of enabling genes to pass to the next generation, and the effectiveness of either strategy is dependent upon the riskiness of the environment; maximizing current reproductive success is thought to be more effective when the risk is high.

British psychologist Daniel Nettle (Nettle, 2005, 2009) suggests that personality types might also represent alternative reproductive strategies. For example, extraverts are more outgoing socially and are also more prone to taking risks compared to their more introverted counterparts. Nettle suggests that extraversion could represent a strategy of maximizing current reproductive success and introversion one of maximizing future reproductive success. The claim is that the more gregarious extraverts are likely to have more children as a result of a greater number of sexual partners, and that their increased risk taking will potentially lead to their accruing more resources (e.g., ancestrally, more success with hunting or, in the modern world, better job prospects). This is a high-risk strategy as the downside of risk taking could be death or injury, or, in modern times, bankruptcy.

In a study of 545 adults Nettle (2005) found that extraversion was positively correlated with the number of sexual partners throughout the lifetime, the number of children borne by other partners and the number of extramarital relationships (they did not necessarily have more children but that is likely to be the result of contraception, something that did not exist in ancestral times). On the negative side, extraversion was also positively correlated with the number of hospitalizations, and extraverts were more likely to have visited the doctor four or more times in the past 2 years. (Because extraversion is a personality dimension, introversion was obviously negatively correlated with all of these.)

So, this provides some evidence that extraverts act to maximize the quantity of offspring that they have at the possible expense of their ability to care for them. When summed across a population, neither introversion or extraversion wins out, hence, both are maintained in the population. The data, however, are somewhat ambiguous. Although extraverts are more likely to suffer injury, they are no more likely to die prematurely than introverts, although this might possibly be due to modern medicine. More seriously, perhaps, there is no evidence that extraverts spend less time engaged in childcare, as the theory might predict (Nettle, 2005).

Niche fitting and frequency dependency

Sometimes the effectiveness of a particular strategy is dependent not just on what you do, but on how that fits in what everyone else does

A nice example of this comes from Steven Pinker (Pinker, 2002). Imagine that there are two routes that you can take in order to drive to work. One of these routes takes the back roads and the other takes the highway. Initially, you take the highway, which is more direct, but you quickly realize that at rush hour the extra traffic makes the journey time unacceptably long. So, instead, you take the scenic route, which, initially at least, takes considerably less time. However, other people discover "your" route and start to take it too, with the result that now the back roads are gridlocked. This has the effect of easing the congestion on the main road, which now becomes the faster route. This principle is in evolutionary terms known as **frequency dependent selection**. In such situations there is no globally optimum best choice; you need to take into account what other people are doing as well. Sometimes, if everybody is using one strategy, it pays you to be different – hence, frequency dependent selection can maintain relatively rare strategies in a population.

> **KEY TERM**
>
> **Frequency dependent selection** The principle in evolutionary theory that states that the effectiveness of a given phenotype – whether or not it is selected – is dependent upon the nature of other phenotypes within a population.

KEY CONCEPTS BOX 10.3

Evolution and individual differences in intelligence – Why do we vary so much?

In addition to personality, psychologists who consider individual differences are also interested in intelligence. Intelligence is notoriously difficult to define, but most psychologists would subscribe to the view that it encompasses the mental abilities of reasoning, planning and problem solving. Most evolutionists interested in the relationship between evolution and intelligence have focused on why our level of intellect is superior to other species (see Chapter 2) rather than why we appear to vary in this attribute so much. Some, however, have considered this variability and have even related it to an evolutionary perspective. Howard Gardner has argued that what intelligence tests measure such as mathematical and linguistic ability is only a small subsection of what intelligence really consists of. To Gardner (2010), the concept of what constitutes intelligence, which he calls **multiple intelligences** (MI), should be broadened out considerably into eight separate areas of human ability:

1. Linguistic (ability to use language)
2. Logical–mathematical (ability to reason mathematically)
3. Visuospatial (ability to manipulate objects in space and mental visualization)
4. Musical (perception of and ability to produce music)
5. Bodily–kinesthetic (ability to control body movements)
6. Interpersonal (ability to understand others)
7. Intrapersonal (ability to understand oneself)
8. Naturalistic (ability to understand and read the natural world).

Of these, only the first three can be considered scholastic intelligence that academic methods, such as IQ tests, measure. Moreover, because, according to Gardner, these eight abilities are largely independent, many people who are seen as less intelligent within our Western academic framework may well show levels of intellect in other areas that would have compensated for their shortcomings during our evolutionary past. If we consider, for example, "interpersonal abilities" then the ability to read others emotional states and respond to them appropriately may well have had inclusive fitness consequences during our ancestral past. Similarly, having superior "bodily–kinesthetic intelligence" may well prove useful in, for example, throwing a spear or dodging a predator. By considering the challenges that our ancient ancestors faced, according to Gardner, we may be able to develop a much broader understanding of what intelligence really is.

Gardner's MI is certainly a novel way of understanding intelligence. It is also one that might help us to understand why we all appear to vary so much in it (i.e., we are only measuring a small proportion of what people are intellectually capable of). While many researchers are sympathetic to this expanded view of intelligence, currently there is little evidence to support the notion that people who are low on what is considered academic intelligence may have compensatory abilities elsewhere (Cooper, 2012).

A more simple view is that the variability that we perceive in levels of intelligence is more apparent that real. This means that compared to the other great apes, we are all very intelligent and the differences we perceive between people today is actually of very little importance in comparison.

Finally, a third, arguably even simpler, explanation is that there might not be an evolutionary explanation for this variability at all – but rather it is simply the outcome of the variability that is caused by the fact that intelligence, being a polygenic trait, varies around the mean as do other polygenic traits, such as height and hair color.

Niche fitting and psychopathy

Psychologist Linda Mealey discusses frequency dependency in relation to psychopaths. **Psychopathy** is a kind of personality, albeit a disordered one, in which individuals have a heightened sense of self-worth, grandiosity, superficial charm, a parasitic lifestyle and engage in manipulative behavior to others. Additionally, they seem to have impairments in many of their emotions – shame, guilt, empathy – that keep most of us on the straight and narrow (Mealey, 1995; Hare, 2003). As a result of their disordered personalities, psychopaths tend to engage in anti-social behavior, cheating, lying, emotional and physical violence, and it has been estimated that 50% of all crime in the United States is committed by psychopathic personalities (Hare, 1993). In short, they are not nice people. Psychopathy is heritable, with genes accounting for around 50% of the variation in the disorder (Larsson, Andershed & Lichtenstein, 2006). Although many psychopaths end up in prison (and it is estimated that they make up 20% of the US prison population; Hare, 1993), many can profit from their parasitic activities. A 2006 book by Paul Babiak and Robert Hare called *Snakes in Suits: When Psychopaths Go to Work* details how many of the "qualities" of psychopathy are similar to those prized in high-level business management, and as a result, contrary to the popular stereotype of the lone serial killer, many psychopaths are in highly paid jobs. Equally importantly, given that this is a textbook on evolution, psychopaths – the majority of whom are male – typically engage in a lot of sex with a lot of people and are thus likely to pass on their "psychopathic genes" to their offspring. They don't usually make good parents, so the psychopath's reproductive strategy is to maximize current fitness.

> **KEY TERM**
>
> **Psychopathy** A clinical condition where individuals are prone to engage in exploitative relationships with others. The **DSM-5** (see Chapter 8) does not have a diagnostic category of psychopathy; instead, it uses the related diagnosis of "anti-social personality disorder" (ASPD) which is a broader definition. Around 4% of men are classed as suffering from ASPD, many see psychopathy as a subset of ASPD (see Mealey, 1995)
>
> Psychopathy affects approximately 1% of the male population (Hare, 1993) and, given the above, one might wonder why this number isn't higher? Mealey suggests that the answer might be because of frequency dependent selection. She argues the key to the success of psychopaths lies in their rarity – unless you work in a prison you are unlikely to meet them very often, meaning that if you do meet them, your defenses are likely to be down and you may end up trusting one of them with potentially disastrous consequences. If, on the other hand, their number were to rise to, say, 10% of the population, then meeting one would become more commonplace and you would therefore become more wary of trusting anyone. This increased vigilance to exploitation would make it more difficult for psychopaths to win your confidence and exploit you. So, the more common psychopathy becomes, the less effective it is as an exploitative strategy. This is, Mealey suggests, what keeps their numbers relatively low.

Niche fitting and non-disordered personalities

Niche fitting might explain the existence of other types of personality, in addition to disordered ones such as psychopathy. As we know (see also Chapters 2 and 3), humans live in groups and have done so for much, if not all, of their evolutionary history. It is possible that groups benefit from having individuals of different personality types. For example, risk takers may make better hunters than the risk averse, introverts who tend to have higher boredom thresholds may make better craftsmen, people higher in neuroticism may be more vigilant to potential threats from outside. We must be clear that we are not here relying on an explanation based on group selection (see Chapters 2 and 11). We are arguing that once we live in groups, then our survival depends – to a large extent – on the way that group functions. If we live in a highly functional group, then we are more likely to survive and reproduce than if we lived in a less functional group. In short, the diversity afforded by having different personality types may have the effect of making a group more functional by enabling people to more naturally fit into roles, to the benefit of the individuals and the genes that they contain.

DISCUSS AND DEBATE

Go through each of the Big Five personality factors. What specific social roles do you think might be appropriate for the extreme ends of each continuum? For example, agreeable/disagreeable, neurotic/stable.

Summary

- Personality theory is a big topic within psychology, and methods such as factor analysis have enabled researchers to measure (to a greater or lesser extent) the personality types of individuals using questionnaires.
- Researchers differ as to how many personality types or factors there are, some such as Eysenck suggesting that there are three, others suggesting that there are five or more. Nowadays, the majority or people that research personality tend to use the Big Five.
- Research on twins suggests that most personality factors are about .5, or 50%, heritable, which means that genetics explains about half of the variation across people.
- Researchers are increasingly studying individual differences in non-human animals such as birds and fish. Perhaps, surprisingly, such animals seem to have something that in humans we might call *personality*. Many of these

differences are heritable, and some of them can be seen to be adaptive.

- A number of theories have attempted to explain why – at the evolutionary level – individual differences exist in humans. Some theories follow the results of non-human animal research and suggest it is down to changing environment, others suggest that it is the result of us trying to fit into different ecological niches, while still others argue that it is merely the result of "noise" in the system.

- There is some evidence that genes are involved in the development of both personality and intelligence. Behavioral geneticists study the degree to which differences between people can be attributed to difference between them in their genes. Genes, that "code for" phenotypic effects are always contingent on the environment. Changing the environment is likely to change the effects of the genes when it comes to intelligence and personality.

- Some experts, including Howard Gardner, consider that the concept of what constitutes intelligence, should be broadened out considerably into a number of aspects of human abilities. Such a view is known as multiple intelligences.

- Evolutionary psychology has only just begun to apply itself to the topic of individual differences in personality, and as a result good, hard evidence for the potential evolutionary benefits is rather thin on the ground at the moment. Based on the recent research on animal personalities where there seems to be good evidence for benefits it seems unlikely that personality differences are just "noise" as a result of sexual recombination. Whether the benefits are the result of niche fitting, trade offs, or some combination of the two is not yet clear.

FURTHER READING

Buss, D. M., & Hawley, P. (Eds.) (2011) *The evolution of personality and individual differences.* Oxford: Oxford University Press. Multi-authored book that considers a wide range of topics relating evolution to individual differences including behavioral genetics, personality psychology and life history theory.

Cooper, C. (2012) *Individual differences and personality* (3rd ed.). London: Hodder Arnold.

Nettle, D. (2009) *Evolution and genetics for psychology.* Oxford: Oxford University Press.

Putting it together

Criticisms, debates and future directions

11

What this chapter will teach you

- Why evolutionary psychologists generally reject the notion of applying evolutionary principles to help shape society.

- There is an ongoing debate between evolutionary psychologists and feminist theorists who claim that it reinforces sexist stereotypes.

- There is debate concerning the degree to which men and women differ in their mate choice criteria.

- There are mixed views about the level at which natural selection operates.

- Some critics have accused evolutionary psychology of the charge of genetic determinism.

- The notion that evolutionary psychologists consider humans to be selfish.

- The concept of applied evolutionary psychology.

Having examined a number of subfields of psychology, where evolutionary theory has made an impact, from developmental, cognitive and social to individual and cultural differences, it is clear that evolutionary psychology has its fair share of critics. In fact as science writer Robert Wright has put it, there appears to be an "anti-ev-psych market niche." Many of these criticisms are related to misunderstandings of what evolutionary psychology entails. In this final chapter we consider the criticisms (whether misunderstandings or potentially valid) that have been leveled at evolutionary psychology and examine current debates within the fraternity. Such criticisms, misunderstandings and debates include sexism, sex differences in mate choice behavior, the notion of the survival of the species, genetic determinism, and the argument that evolution is no longer important to our species since we have developed culture. As is apparent these themes largely arise out of areas of academic interest that we have presented in various preceding chapters.

In order to consider these debates, we present them as frequently asked questions. We begin with a question that has regularly been asked – should evolutionary psychology be used to shape society?

Should evolutionary psychology be used to shape society?

There has been a long history of people using evolutionary theory to help better society. From the Social Darwinism of Herbert Spencer in the 19th century to the eugenics movement, which argued that we should encourage people with desirable traits to breed with one another and discourage those with undesirable traits from breeding (or in extreme cases, sterilize them).

Attempts to use Darwinian theory to reform society usually have at their center the belief that society is dysfunctional because people are being expected to do things that they were not evolutionarily designed to do. Some of these are plausible. For example, shift work is quite unnatural in this regard. Humans are a diurnal species (active during the day) and having people work through the night is associated with many negative health consequences (Harrington, 2001). So, in a small way, understanding something about evolution could help us to redesign aspects of society to the betterment of everyone. But, we must be careful not to overextend this line of reasoning, because we are in danger of committing the **naturalistic fallacy** (see also Chapters 4 and 9), which states that if something is natural, then it must be good. This is a genuine fallacy – many things such as radiation and snake venom are natural, but that does not make them good things.

The naturalistic fallacy is at its most dangerous, however, when it is applied to moral or ethical debates. Take aggression. From an evolutionary perspective it is to be expected that under certain circumstances aggression can be a good thing from the point of view of the aggressor. For example, if an individual or its offspring are threatened, aggression can be used as a form of defense. It may also be used non-defensively, for example, in predation or in order to steal food from another person. In both situations aggression benefits the aggressor – we may even say that aggression is "natural" – but does that make it morally justified to use violence?

As an analogy, imagine that someone owned a pet dog. At some point the dog gets off its leash and kills a child, resulting in the dog's owner appearing in court. In his defense the owner tells the jury that in its ancestral environment a dog is a natural predator – he even pulls out diagrams showing brain regions that are specifically designed by evolution for predation – and as such, he argues, it is understandable that the dog killed the child. What would you think of such a defense? What would the jury think? What would the judge say? Most likely the reply would be that one of the responsibilities of dog ownership is to ensure that the dog is sufficiently trained and socialized to minimize the chances of such an attack. The judge may also point to evidence that the vast majority of dogs behave in an acceptable manner (this is the reason we permit dogs to be walked in parks but not tigers). In the same way that you cannot exonerate poor behavior on the part of a dog by citing evolutionary evidence, you cannot justify your own poor behavior by suggesting that it is natural. Part of participating in society is that we permit ourselves to be likewise domesticated and curb our "natural" instincts.

Some recent research gives the lie to the fact that "natural" is better. Figure 11.1 shows that the chances of dying as a result of warfare are far higher for the presumably more natural non-state societies that it is for state societies (see Pinker, 2012). This is not the result of any evolutionary change but rather the civilizing effect of apparently unnatural social contracts that exist within states: police forces, the rules of law and so on.

So, to sum up, evolutionary psychology may be able to help design a better society, but in a small way. Evolution itself provides us with little or no guidance as to how a fair society should be run because it provides us with no absolute moral principles. We must bear in mind that what might "naturally" benefit one section of society might – equally naturally – harm another section.

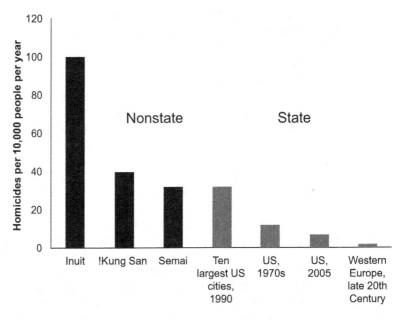

Figure 11.1 Homicide rates for state and non-state societies.

Source: Adapted from Pinker (2011).

DISCUSS AND DEBATE – SHOULD RESEARCH DEMONSTRATING DIFFERENCES BETWEEN THE SEXES OR BETWEEN GROUPS OF DIFFERING ETHNICITY BE FREELY PUBLISHED?

Relating to the point discussed above, what should be the limits on published research? Should research showing that the sexes (or races) differ intellectually be published even if some people might use this research for undesirable purposes (such as justifying racist or sexist policies).

Is evolutionary psychology sexist?

A number of feminist theorists have taken issue with evolutionary psychology on the grounds that it appears to reinforce sexual stereotypes. Common feminist criticisms are that evolutionary psychologists present an inflexible view of gender roles and that this view favors men over women: [E]volutionary psychology has come to rely on assumptions deemphasizing the pliability of psychological mechanisms and the flexibility of human behavior. (Liesen, 2007, p. 51) Feminists reject the current assumption of EP –according to which there is a universal human nature – for many of the same reasons that they have rejected the idea of a universal human nature: the idea of a universal human nature is a fiction that privileged groups generate in order to further justify and buttress that very privilege; the privileged group is never women.

(O'Donavan, 2013)

Evolutionary psychologists do consider that some gender differences have arisen from **sexual selection** acting differentially on the two sexes. This means that in some areas men and women have faced different recurrent adaptive problems – such as hunting and gathering – and in mate choice criteria and levels of aggression (see below). We should also note, however, that in most areas men and women will have faced the same recurrent adaptive challenges – such as avoiding predators and parasites, securing sufficient warmth and protective surroundings and engaging in social exchange.

To recap – sexual selection concerns gaining access to the opposite sex either directly by impressing them, by being "sexy" (e.g., gaudy plumage of the birds of paradise – intersexual competition) or by competing with the same sex (e.g., bulls are big and strong and have large horns – intrasexual competition). Remember that **parental investment** theory is the notion that the greater investing sex is the one that is more choosy and the lesser investing sex is the one that competes for the attention of the greater investing sex. In most vertebrates females are the greater investing sex (see Chapter 4 for some exceptions), and thus females tend to be the choosier ones and males compete for female attention.

Females, for their part, are choosy about which male they will mate with because, due to their higher level investment, they have more to lose if they make a bad mate choice. Hence, asymmetrical parental investment leads to what has come to be known as the **males-compete/females-choose (MCFC)** model of male and female sexual psychology (see later).

Two examples of important publications that examined the notion of evolved sex differences are those of Buss (1989 – see Chapter 7) and Archer (2009 – see Chapter 3). In David Buss's (1989) empirical study of mate choice preferences it was reported that women, more so than men, rate status and good financial prospects highly in a romantic partner and that men, more so than women, rate physical attractiveness highly in such a partner. Buss suggests that this may be traced back to sexual selection acting differentially on males and females. Note that these differences were not driven by stereotypes of gender role but rather by hypotheses derived from sexual selection theory. In the case of differences in the importance of physical attractiveness, Buss predicted this finding because females have a more limited period of fertility, leading to males favoring attractiveness that is directly rated to youthfulness. In the case of women favoring high status in men, this is derived from the fact that females have a higher level of parental investment and hence require signals of investment from men to form a relationship (with high status being a likely signal of the ability to invest). In the case of John Archer's (2009) review article, evidence was presented that indicated sex differences in aggressive behavior are the outcome of sexual selection rather than social learning. Once again, this was derived from sexual selection theory since males generally compete for females, leading to a lowering of their threshold for aggressive response. Archer reports that both that levels of physical aggression are more common in men than in women across studies and that sexual selection theory provides a robust explanation for such differences. If uncovering differences between the sexes is your definition of sexism, then evolutionary psychology is indeed sexist. But to claim this would be, in effect, to claim that the natural world is sexist. This would be a strange way of defining sexism. Hence, we feel that evolutionary psychologists should not be criticized for uncovering differences between the sexes that are driven by sexual selection theory. Moreover, perceiving a difference does not mean that one sex is considered superior to the other. Are ova inferior to spermatozoa? Is having more testosterone superior to having more estrogen? Such questions are meaningless.

We need also to consider the more specific criticism that evolutionary psychologists present an inflexible view of gender roles. The sex differences in mate choice preference and in levels of aggression reported above appear to be both robust and universal (Archer, 2009; Buss, 2014). Despite this, it is clear that, even in areas where there are well-documented differences, male and female behavioral responses overlap greatly with, for example, some females behaving more aggressively and **promiscuously** than most males and some males behaving in a placid way and entering into enduring **monogamous** relationships. It is important to realize that we are concerned here with average differences between the sexes and that such differences may be increased or decreased by the expectations of a given society. Interestingly, there is no evidence that these sex roles are reversed in any society (Brown, 1991). Evolutionists do not stress inflexibility in individual behaviors but rather point out that these general differences in mean scores are conducive with sexual selection theory. In fact, the portrayal of evolutionary psychologists presenting an inflexible view of human sex differences in behavior and performance is not substantiated by a review of the literature (Vandermassen, 2005 – see below). As David Buss and David Schmitt have pointed out, the evolutionary psychology literature repeatedly describes human behavior as adaptively flexible (Buss & Schmitt, 2011).

There is an additional danger that many of the feminist critiques of evolutionary psychology may commit the **moralistic fallacy,** which is the polar opposite to the naturalistic fallacy discussed above. The moralistic fallacy is the assumption that because we view something as good or desirable it must therefore be true, or that something must be false if we disagree with it (see Chapter 4). Feminism has been extremely important as a critique of male patriarchy and has been of prime importance in promoting a more equal society. And it doubtless has a role to play in evolutionary psychology in challenging potential patriarchal interpretations or assumptions. But it makes no sense to take the further step of rejecting data or theory just because you don't agree with its implications. One can certainly see why some feminists would wish to dismiss the idea of an evolved human nature, it would be so much easier to argue for equality if deep down we were all the same. But, as we discussed in the first section of this chapter, evolved differences should not be seen as compromising equality, because one should not base political or moral views on evolutionary theory.

In an important and balanced review of feminist criticism of evolutionary psychology text called *Who's Afraid of Charles Darwin? Debating Feminism and Evolutionary Theory*, Griet Vandermassen (2005) concluded that feminist scholars had repeatedly mischaracterized evolutionary psychology (supported by Winegard, Winegard & Deaner, 2014). Additionally, however, she also criticized evolutionary psychologists for not taking into account well-developed feminist theory. More recently, Alice Eagly and Wendy Wood (2011), in reviewing the relationship between evolutionary psychologists and feminist theorists, found that a great deal of mutual mistrust continues. Clearly, the marriage between feminist theorists and evolutionary psychologists has yet to be consummated.

Why do men and women differ in their mate choice criteria?

A separate but related question to the notion of evolutionary psychology being sexist is the question as to whether or not males and females differ in their mate choice criteria. As we pointed out, evolutionary psychologists explain current sex differences in behavior (such as mate choice preferences) by applying Darwin's theory of sexual selection and Trivers's theory of asymmetrical parental investment. As we saw earlier on, evolutionary psychologists generally subscribe to the males-compete/females-choose (MCFC) model and consider that current sex differences in mate choice strategies, while influenced by culture, can ultimately be traced back to this asymmetry in investment.

Social psychologists Alice Eagly and Wendy Wood (mentioned earlier) are not entirely convinced by the explanation of sex differences presented by evolutionary psychologists and instead subscribe to the view that social learning plays a prominent role in our development of sex roles. Their initial view, which they called the **social role theory** of sex differences, relied heavily on learning from others about how each sex is expected to act rather than suggesting there are evolved psychological differences that lead to this dimorphism in human behavior (Eagly, 1987; Eagly & Wood, 1991; see Chapter 3). In this way males, for example, learn that it is more permissible to engage in rough-and-tumble play as children and, in adulthood, to seek casual sex. Females, in contrast, are dissuaded by their elders from either of these forms of behavior. Eagly and Wood have also argued that a preference for males with resources is directly related to the fact that in many cultures women have restricted access to financial resources. Hence, rather than having an evolved disposition toward such men, they are forced to favor them due to their lack of direct access to financial rewards (Eagly & Wood, 1999).

> ### KEY TERM
>
> **Biosocial construction theory of sex differences**
> A theory, derived from social role theory, that considers that biological factors interact with social ones in the development of male and female sex roles.

Although broadly critical of evolutionary psychology, Eagly and Wood later began to accept that evolved characteristics do play some sort of role in such differences, renaming their model as the **biosocial construction theory**. The biosocial construction theory explains sex differences as arising out of an interaction between social learning and biologically evolved differences – but these biological differences are mainly physical with psychological differences following as a secondary consequence (Eagly & Wood, 2011; Wood & Eagly, 2012). An example of this would be that men, finding themselves to be generally a bit taller, faster and stronger than women, then, as a response to these physical differences, gravitate towards typical masculine social roles. Hence, Eagly and Wood open the door to evolution in as much as physical differences are considered to lead to psychological ones – but they reject the general view of evolutionary psychologists that the sexes have inherited psychological differences (see Chapter 3). There are still debates concerning to what extent psychological sex differences arise as a direct result of differing pressures during our evolutionary past or are due to feedback from physical differences (Laland & Brown, 2011). In either event, as we saw above, evolutionary psychologists do not see such differences are hardwired, and it is generally accepted that learning, albeit channeled learning, is import in all human responses (Buss & Schmitt, 2011).

Darwin called differences in structure and behavior between the sexes **sexual dimorphism**. As outlined above, to evolutionary psychologists such sexual dimorphism arose out of sexual selection and the fact that females invest more heavily in offspring than males. It is easy to fall into the trap of considering some species as dimorphic and others as monomorphic (that is no physical differences beside the reproductive organs with no clear behavioral differences). This conception is a gross simplification, however. In reality, species vary along a spectrum in their degree of sexual dimorphism. At one end of the spectrum, elephant seal males are around four times the size of females – with the former having a very low threshold for aggressive response when compared to the latter. In contrast, in the European robin, both sexes look and behave so alike that males often chase receptive females from their territories when they first make an approach during the breeding season (they clearly find it difficult to tell the sexes apart – but somehow eventually get the message!).

Recently, the "males-compete/females-choose" model of sexual psychology for humans has been challenged from within the field of evolutionary psychology. Steve Stewart-Williams and Andrew Thomas, of Swansea University, have suggested that, because human offspring are born so immature and therefore require aid from both parents, humans are in the main a pair bonded species where, due to more equal investment, the sexes are more similar than different. Stewart-Williams and Thomas (2013a, 2013b) suggest that evolutionary psychologists have overemphasized human sex differences and that, in their view, we are closer to being monomorphic than dimorphic (more robin-like than elephant-seal–like). Stewart-Williams and Thomas call their view of human sexual psychology the **mutual mate choice model (MMC)**. To back up their assertion they suggest that while there are measurable sex differences in men and women's willingness to engage in casual sex, this difference is not as large as is generally assumed. Many studies have found a statistically significant difference here, with males being far more willing than females (see, for example, Buss & Schmitt, 2011). Stewart-Williams and Thomas point out, however, that while the differences uncovered are robust they are not large. Hence,

Table 11.1 Difference between males-compete/females-choose and mutual mate choice models of sex differences in mate choice behavior.

Males-compete/females-choose (MCFC)	Mutual mate choice (MMC)
• Overall, sexual dimorphism in behavior is high.	• Overall, sexual dimorphism in behavior is relatively moderate.
• Males very competitive for mates.	• Both sexes are choosy about long-term mates.
• Females very choosy about mates.	• Both sexes quite competitive for mates.
• Large difference between sexes in level of in parental investment (females invest more heavily in offspring).	• Small difference between sexes in level of in parental investment (because males contribute to raising offspring).
• Is the most commonly accepted model of sex differences in evolutionary psychology.	• Is not currently frequently accepted in evolutionary psychology.

many relatively small studies reporting a small sex difference does not add up to a large study reporting a large difference. Stewart-Williams and Thomas make the same point with regard to sex differences in levels of aggression and risk taking (i.e., differences between the sexes in, for example, rates of violent crime) that they have also been exaggerated. Table 11.1 outlines the differences between MCFC and MMC models of sex differences in mate choice behavior.

Some evolutionists have been sympathetic to the MMC model (Buss, 2013b; Eastwick, 2013; Miller, 2013), but others have been critical (Campbell, 2013a; Kenrick, 2013; Roberts & Havlíček, 2013). It is clear that this debate is not close to being put to bed, but even to those who don't agree that we are more monomorphic than dimorphic, it has at least reminded evolutionary psychologists that answers to such questions are not cut and dried. One possible solution to the MCFC vs. MMC debate is that we are both, in as much as humans are notoriously flexible in their reproductive behavior. Under some circumstances, men may form enduring pair bonds and invest heavily in offspring – meaning that in such relationships behavioral dimorphism is low. Under other circumstances, men may chose to engage in a series of short-term romantic liaisons providing little or no parental care (perhaps when women outnumber men, such as during wartime). Such a relationship would then be seen as relatively sexually dimorphic (Betzig, 2013). Hence, both MMC and MCFC models of behavior might exist within the same population and even within the same individuals, under different circumstances.

"Survival of the species" – What does natural selection select for?

When people consider the term *survival of the fittest* they often assume that this refers to survival of the fittest species. We have regularly read textbooks that state explicitly that natural selection promotes survival of the species. To modern-day evolutionists, however, this phrase is problematic. Imagine a species where individuals behave in ways that promote the survival of the species. This means that when animals forage for food it does not matter whether they then consume it themselves or pass it on to other (non-related) members of the species. Moreover, individuals would be just as likely to help others reproduce as to do so themselves. If an animal's behavior is designed by natural selection to help the species survive, then it matters not a jot whether individuals help themselves or other (non-relatives) in these aims. Logically, however, individuals that behave in this way are likely to be rapidly replaced by other individuals who favor their own feeding and reproductive output (Maynard-Smith, 1964). Hence, much of natural selection involves competition for resources and reproductive output between members of a species. This is why we should not refer to natural selection as acting for the survival of the species. Note that if the species survives this is not due to natural selection acting for the survival of the species but is a secondary consequence of the self-interested behavior of its members.

Most evolutionary psychologists subscribe to the view that individuals behave in ways that aid their own survival – or that of close relatives that share many genes by common descent with them (see later). Hence, as we discovered in Chapter 1, most evolutionary psychologists see the gene as the unit that natural selection "selects" (the "genes eye view," Dawkins, 1976/2006; see Chapter 1). It is important to realize, however, that while all modern-day evolutionists disavow the notion of survival of the species, not all subscribe to the view that the gene is the *only* level at which selection operates (Laland & Brown, 2011).

Some evolutionists consider that natural selection can operate at levels above the gene. In particular, a number of evolutionists, critical of the genes-eye-view of evolution have, over the last 20 years, developed a **multilevel selection** model. Multilevel selection posits that natural selection can select at a number of different levels or "units" including the group, the individual and the gene. Such

KEY TERM

Multilevel selection The notion that natural selection acts at a number of different levels, including the gene, individual, group and population.

a view has been expounded by American evolutionists David Sloan Wilson and Elliot Sober (Wilson & Sober, 1994; Sober & Wilson, 1999). Sober and Wilson propose that under some circumstances, as well as genes and individuals being the unit of selection, groups can compete with each other in terms of reproductive output. Wilson and Sober consider that natural selection acts at these various levels of competition rather like a Russian doll, with each layer of competition embedded inside the next level up. Wilson and Sober argue that members of a particular group can outcompete members of another group when in competition for resources and hence we can think of the group as the unit of selection under some circumstances.

We can well imagine scenarios such as warfare where an entire group can thrive at the expense of another group. As Steven Pinker (2012) has pointed out, however, "natural selection selects for a gene if it causes a behavior that leads to that gene increasing in frequency in the population, not some other arbitrarily defined scale such as social partners." The big problem with MLS is, while it may in principle be possible for selection to occur at the individual or group level under certain circumstances, there are very few (if any) instances where this is necessary. Hence, to us, and most other evolutionary psychologists, multilevel selection can ultimately be distilled down to selection at the level of the gene.

Is evolutionary psychology based on genetic determinism?

Leading on from the last section, we can think of the gene as the currency of evolution by natural selection. Because this is the view of mainstream evolutionary psychology and because we relate evolutionary processes to current human behavior and internal states, then the approach is regularly criticized for supporting **genetic determinism**. *Genetic determinism* is the notion that human behavior is controlled by genes and hence our destiny is written in our genetic code. The portrayal by its critics that evolutionary psychology as being based on genetic determinism is widespread and enduring (see, for example, Lickliter & Honeycutt, 2003; Smith & Thelen, 2003; Winegard et al., 2014). In one sense this assumption is unsurprising since, to evolutionists, the success of a given behavior is ultimately measured in terms of inclusive fitness (i.e., the proportion of genes passed on to future generations), or, rather, how that behavioral response would have boosted inclusive fitness during ancestral times. There are, however, two problems with this view. First, no evolutionary psychologist has ever (to our knowledge) suggested that genes *determine* behavior, and second, it would not make sense to have behavior genetically determined. This is the case even in simple organisms. All organisms that are capable of behavior have to show some flexibility in their responses, and hence, behavior, while being influenced by genes, cannot be entirely genetically determined. A fixed, genetically determined response is a recipe for extinction. Think about it – genetically determined behavior by definition shows no flexibility.

In the words of British biological psychologist Fred Toates, "evolution cannot guarantee a correct response every time" (2011, p. 312). This means that even simple animals demonstrate the flexibility in behavior that we call *learning.* Take the marine sea slug, *Aplysia,* as an example (see Figure 11.2). *Aplysia*, which belongs to the soft-bodied animal group known as mollusks, has a limited number of **neurons** in its nervous system. Despite this, it is capable of learning to alter its responses to external stimuli. The development of these neurons is very much guided by its genome. When touched, *Aplysia* withdraws its gill and siphon (retractable breathing and waste organs, respectively). This

Figure 11.2 *Aplysia californica,* the Californian sea slug.
Source: Image copyright © Elliotte Rusty Harold/Shutterstock.com.

withdrawal reflex is as good an example of an instinctive response as you will find in the animal kingdom (Kandel, 1976; Toates, 2014). This means that it is predictable and demonstrates a stereotypical pattern (i.e., all members of the species withdraw their gill and siphon in the same way). When, however, it is repeatedly touched, after a while it no longer withdraws these appendages (Kandel, 1976). The sea slug has learned that the touching poses no threat and modifies its response to reflect this discovery. A strict adherence to genetically determined behavior would have the sea slug producing an endless series of inappropriate responses to a harmless stimulus – a response that clearly would be a poor adaptation and would very rapidly be removed from a population by natural selection.

If an animal with as simple a nervous system as a sea slug can demonstrate flexibility of behavior, then how much more flexibility can we expect from animals as complex as a big-brained primate? There is a world of difference between the genetic determinism that evolutionary psychologists are charged with and their actual view that genes influence behavior. Proposing that humans develop a brain that is influenced by genes to support flexible responses that would have aided survival and reproduction under ancestral ecological and social pressures is a far cry from genetic determinism. In the words of David Buss and David Schmitt (2011, p. 769):

> Human behavior is not and cannot be, "genetically determined"; environmental input is necessary at each and every step in the causal chain – from the moment of conception through ontogeny and through immediate contextual input – in order to explain actual behavior.

Does evolutionary psychology argue we are all selfish?

A common misunderstanding of evolution in general and evolutionary psychology in particular is that it inevitably emphasizes selfishness. While it is true that evolutionists are apt to discuss "individuals acting in their own interests" (see the section on levels of selection), it is not inevitable that selfishness happens as a matter of course. At least one reason (and there are many) why this misconception still holds sway is down to Richard Dawkins's (1976/2006) book *The Selfish Gene,* the title of which contains the pernicious "s" word. As Dawkins himself has pointed out, the book is about selfishness at the level of the gene, not about selfishness at the level of the individual, and people who believe the latter are probably those who have read the title while ignoring the substantial footnote of the book itself. In fact, the book contains a great deal about how altruism of various forms could evolve; not only that, but the idea of the "selfish gene," stemming from work from researchers such as George Williams, Robert Trivers, John Maynard Smith and (especially) William Hamilton is able to explain why some organisms are not selfish.

We discussed this in Chapter 1, but it is a point that is always worth emphasizing: genes may be "selfish," but this does not mean that individuals are. And humans most certainly are not selfish. At least not always. For example, humans have the emotional facility known as love. It may sound unpoetic, but love is an emotion of selflessness: it encourages individuals to act in the interests of others – whether the "others" be our children, our partners or our best friends. It is easy to see why our genes should make us concerned about our children: any gene that leads an individual to promote the well-being of its offspring is likely to benefit because the offspring is likely to contain copies of itself (this is Hamilton's rule, explained fully in Chapter 5). Kin altruism is not restricted to offspring, selfless acts to any reasonably close relative (siblings, cousins, nieces, nephews) can potentially evolve so long as the benefit to the recipient is greater than the cost to the actor multiplied by their relatedness (see Chapter 5).

What Hamilton's rule shows is that genes acting in their own interests can produce individuals that act in the interests of others. Such selflessness is not restricted to "higher" animals. Many animals from a range of taxa: vertebrates, mollusks, arthropods and even single-celled slime molds show altruistic behavior to relatives. In fact, the more we look at nature the more it seems that cooperation among kin is the rule rather than the exception, and the more we consider human physiology and psychology, the more it becomes clear how selfless humans are as well.

Consider your sexual organs: whom do they benefit? They may provide you with a fleeting pleasure, but ultimately your sexual organs are structures that enable genes to leave your body and recombine with the genes of someone else. If you happen to be female, then likewise your breasts are (largely) structures that your genes have provided to sustain their own existence via lactation. And both males and females typically engage in childcare to a greater or lesser extent. Why? Because of the aforementioned love. A purely selfish animal would have no use for love. Not only love, but also shame, guilt, compassion, contempt and certain forms of punitive anger, all of which exist to motivate and manage cooperative behavior. In humans, in particular, cooperation need not be among kin, and it is likely that we have evolved mechanisms such as reciprocity (Trivers, 1972; see Chapter 5) which involve many of the aforementioned emotions.

We are not, of course, *unconditionally* selfless. Individuals of many species will engage in infanticide (the killing of young offspring) if the offspring threaten their life as a result of competition between parent and offspring over limited resources (as well as other reasons – see Hausfater & Hrdy, 1984). And – as we discussed in the first section – if we are threatened or feel cheated we might well retaliate violently, but the view that nature is entirely "red in tooth and claw" is rapidly changing as a result of the work of evolutionists of all kinds.

Is evolution unimportant now that we have culture?

The view that culture somehow renders evolutionary explanations of behavior unimportant is one that we have encountered many times as we discuss evolutionary theory with our colleagues. Hopefully Chapter 9 will have shown that this is simply false. Partly this belief comes from the nature versus nurture or culture versus biology controversies which are fond of lumping the many forces that shape who we are into one or other category and then asking which is correct. A barely more sophisticated version accepts that both matter, but then loses the plot by asking which one is most important. The upshot of this faulty thinking is that when we emphasize the importance of evolution on human behavior we are often misconstrued as believing that culture is somehow unimportant. This is also false. Culture is of course a very powerful influence on how we behave (again, see Chapter 9 for some examples), but this does not diminish the importance of evolution. In fact, many theorists see culture itself as being something that requires specially evolved mental hardware (the ability and desire to imitate, for example) in order for it to arise. In this sense culture isn't something that just exists independent of humans. The human brain is as much a cause of culture as it is a product of culture.

The future of evolutionary psychology

Evolutionary psychology has come a long way during its brief but eventful history. From the early work by Cosmides and Tooby on cheat detection in the late 1980s and early 1990s (see Chapter 7), it has expanded to examine a wide range of psychological phenomena, from sex differences in mate choice (see Chapter 4), through work on the adaptive nature of emotions and why our cognitive processes demonstrate limitations (see Chapter 7), to ideas based on evolutionary theory as to why individuals and individual cultures might vary so much (see Chapters 9 and 10). What, we might ask, is next for evolutionary psychology? Having developed a large body of theoretical work and a huge amount of lab and field-based data, perhaps the next step will be in a more applied direction? This progress from developing theories and gathering empirical findings to elucidating how such developments might lend insight into applied issues is the typical path for a new subdiscipline in psychology (Roberts, 2011). This raises the question – how might an area that is often portrayed as being on the esoteric end of the psychological sciences be applied to helping people? To put it another way, can knowledge of how past selection pressures are likely to have shaped the minds of our ancient ancestors be of some use today? Perhaps more so than you might think.

Applied evolutionary psychology

In recent years a number of evolutionists have begun to consider how evolutionary psychology can both help us to understand why we develop mental health problems and suggest ways in which they might be treated (see Chapter 8). Alarmingly, rates of depression have risen around the world in the last 70 years, and they have risen most steeply in the rich, "developed" nations of the world. As we saw in Chapter 8, there is now evidence that this rise in depression in the wealthy West is likely to be related to a "mismatch" between our current lifestyle and the conditions under which our ancestors evolved. Moreover, there is growing evidence (albeit currently small scale) that, if we can shift facets of our lifestyle to more closely resemble those of our ancestors, we may then be able to reduce rates of depression (Ilardi et al., 2007; Ilardi, 2010). Given the success of such studies (and their importance to our quality of life), we would predict that findings from evolutionary psychology will begin to become incorporated into therapeutic services over the next 20 years.

DISCUSS AND DEBATE – BACK TO A SIMPLER LIFE?

Ilardi suggests that if we shift our lifestyle to one that more closely mirrors that of our ancestors we could greatly reduce levels of depression. Can you think of any downsides to this argument? Make a list of features of the lives of our savannah-dwelling ancestors that you feel are superior to our lives today and another list of features of modern life that are superior to those of our ancestors. Which life style would you prefer?

Another applied area is the evolutionary psychology of criminal behavior (Figueredo, Gladden & Hohman, 2011). The academic study of criminology is largely a reserve of applied sociologists who therefore take on a proximate and environmental perspective to explain crime (Taylor, 2015). In recent years, a number of evolutionary psychologists have attempted to add the ultimate perspective to help explain such deviant behavior. Drawing on past theory and empirical observations, Figueredo et al. (2011) have outlined how the ultimate perspective might help to explain and possibly reduce a number of different forms of criminal behavior. In the case of psychopathy, an understanding of how psychopaths might manipulate the general human social system of reciprocation before moving on might help criminologists to locate such individuals (see Chapter 8). Furthermore, an understanding of Daly and Wilson's "Cinderella effect" (i.e., higher rates of abuse by stepparents than by biological parents; see Chapter 5) might help to predict more accurately when a child is in danger of abuse. Finally, an understanding of parental investment theory (see Chapters 4 and 5) may help us to understand what sort of social conditions might increase the chances of infanticide occurring.

Evolutionary psychologists are only just beginning to scratch the surface when it comes to applying evolutionary theory to help gain insights into problems associated with mental health and criminal behavior. Many of these early attempts have been criticized (Buller, 2005b), and a number of them may turn out to be blind avenues of research. That is generally the case with science. But, to ignore the potential insights that an ultimate perspective might add to all of the proximate perspectives that have been developed would surely be a missed opportunity. The social sciences have come a long way in understanding why individuals become either mentally ill or prone to criminal behavior, surely understanding why, as a species, we have such propensities is more than a matter of pure academic intrigue.

The future of psychology is evolutionary psychology?

Here we address not so much the future of evolutionary psychology, but the future of psychology itself which, we believe, needs to adopt evolutionary theory if it is to more completely and more properly understand human behavior. This is a bold claim; why do we make it? The first is that evolutionary theory allows us to answer ultimate questions (see Chapter 1), and these cannot be answered by mainstream psychology. Questions such as Why do we (usually) love our children? Why do we find some things disgusting? and Why do men and women generally differ in their sexual behavior and attitudes? In Chapter 1, we saw that the answer to the second of these is that disgust is an emotion that protects us against infection: things that are likely to contain pathogens promote an aversive disgust response, which causes us to avoid them.

A second reason is that unlike traditional psychology it doesn't always place the individual at the center of the action. Even social psychology has traditionally explained social behavior as being for the benefit of the individual. Instead, as we have seen, many of our behaviors are for the benefit of our genes: childcare being the most prominent example. This, as we saw earlier, is known as the **gene-centered** view of life.

Finally, we believe that by adopting an evolutionary framework, psychology can become more integrated into the rest of the life sciences rather than sitting apart with its own unique view of human nature. There is evidence that this is beginning to happen. Over the past 20 years we have started to see a gradual increase in the number of evolutionary psychology research papers being published in mainstream psychology journals rather than just the specialist evolutionary psychology journals. Doubtless, it will take time until evolutionary theory becomes the default framework in psychology, and at least part of the reason for this is that many psychologists have been very poorly trained in evolutionary theory. Hopefully, after reading this book you will be better informed and ready to become part of this new world view.

Summary

- During the late 19th and early 20th centuries, attempts were made to link Darwin's natural selection to help redesign aspects of society (such as who should breed with whom). Modern evolutionists reject such a notion because suggesting that what is natural is not necessarily therefore morally desirable (to make such a link is known as the *naturalistic fallacy*).
- Some feminist theorists have suggested that evolutionary psychology reinforces sexual stereotypes. Evolutionary psychologists have, however, suggested that such a criticism is misplaced because uncovering evidence of sex differences that may be related to Darwin's sexual selection

is a scientific rather than a moral or political process. More-over, to assume that something that is desirable is therefore true commits the moralistic fallacy.

- Evolutionary psychologists have uncovered evidence that men and women exhibit some differences in their predilec-tions for a mate. Social psychologists have suggested this is due to learning about sex typical behavior within a given society – the social role theory of sex differences. A devel-opment of this theory – the biosocial construction theory – suggests that evolutionary processes are involved but that socialization factors are preeminent. Evolutionary psychol-ogists reject these explanations of sex differences in behav-ior, explaining such differences in terms of one of two hypotheses that have been derived from sexual selection theory – the males compete/females-choose (MCFC) and the mutual mate choice (MMC) theories. MCFC suggests that when it comes to choosing mates, males are the competi-tive sex while females are the choosy sex. In contrast, MMC suggest that both sexes are broadly similar in their level of choosiness and competitiveness. Currently, this debate is yet to be resolved.

- Some evolutionists consider that natural selection can oper-ate at multiple levels – favoring the species, population, group, individual or gene. This viewpoint is known as the *multilevel selection model*. Most Evolutionary psychologists reject the notion that we have evolved to aid our group or species survive, considering instead that our behavior has evolved to aid the individual and ultimately the individual's genes.

- Critics of evolutionary psychology have accused its propo-nents of genetic determinism – that is that our behavior is encoded in our genes and not subject to environmental influence. Evolutionary psychologists reject this criticism as they consider that the human brain has evolved to demon-strate adaptively flexible behavior that is influenced, rather than determined, by genes.

- Evolutionary psychologists do not consider that we have evolved to be selfish, but rather, genes which have evolved to enhance copies of themselves, often do so by influencing behavior through altruistic means. This may occur via kin altruism (aiding relatives) or through reciprocation (where individuals take turns in providing aid for each other).

- The rise of human culture does not mean that evolution no longer has any explanatory power. Evolution and culture are perceived as part and parcel of each other and both feed back on each other. The evolved brain allows for culture to develop and culture allows human brains to thrive.
- During the first 20 or so years of its existence as an academic subdiscipline, evolutionary psychology has developed a large body of theory and data. We consider that one important development for the next 20 years will be in an applied direction. By increasing our understanding of the ultimate level of causation, evolutionary psychologists, for example, may be able to develop therapies to aid people with mental health problems.
- Despite its critics, evolutionary psychology has had a growing impact on mainstream psychology in recent years. We consider that the subject can benefit greatly by an integration of the evolutionary approach into all aspects of mainstream psychology.

FURTHER READING

Roberts, S.C. (Ed) (2011) *Applied evolutionary psychology*. Oxford: Oxford University Press.

Workman, L., & Reader, W. (2014) *Evolutionary psychology* (3rd ed.). Cambridge: Cambridge University Press.

Glossary

Affective neuroscience Jaak Panksepp's term, used to describe the neurological bases of emotional experience and expression.

Agentic traits The tendency to make choices and impose them on others.

Allele (full name – allelomorph) The various forms of a gene that can occupy the same position on a chromosome.

Anisogamy A state where the gametes are of different shape and size. Ova, for example are much larger than sperm.

Anthropoid ape "Man-like" primates subdivided into the great apes (chimpanzees, gorillas, bonobos and orangutans) and the lesser apes (gibbons and siamangs).

Anthropomorphism Attributing human-like cognitive and emotional attributes to other species.

Anxiety disorders Serious mental health disorders whereby an individual has an extreme reaction to various social or environmental circumstances. Examples of anxiety disorders include phobias, separation anxiety and social anxiety disorder.

Apomixes The technical term for sex (sexual reproduction).

Autism spectrum disorder (ASD) A range of developmental disorders involving social and communication deficits. Individuals with ASD generally have difficulty forming relationships and engaging in repetitive behavior patterns.

Behavioral genetics The field of study that considers the role of genes in behavior in humans and animals.

Bipedal locomotion Walking upright on two legs.

Bipolar depression Serious mental health disorder whereby an individual has alternating bouts of depression and mania (also known as *bipolar affective disorder* and *manic depression*).

Biosocial theory (also called *biosocial construction theory*) In relation to sex differences in behavior, the view that physical, biologically influenced characteristics have a secondary influence on the development of gender roles.

Black box In science, a device that is viewed in terms of input or output because the inner workings are opaque.

Broaden-and-build A theory to explain the function of positive emotions. This involves expanding our social networks, enriching our knowledge base, and developing more flexible problem-solving skills.

Candidate gene A gene suspected of being involved in a specific condition or disease.

Catalyst/catalyze Chemical that enables, speeds up or slows down a chemical reaction without being used up itself.

C-F continuum The position on a spectrum where an organism falls in relation to devoting its time and energy to maximizing current (C) or future (F) reproductive success.

Chromosome A lengthy string of genes.

Cinderella effect The notion that stepparents tend to lavish less attention on stepchildren than on their own biological offspring, derived from kin selection theory.

Concordance rate A term used in genetics to describe a measure of the likelihood of an individual of a pair (such as twins) having a specific trait if the other individual in the pair has that trait.

Common ancestor An individual that gave rise to two or more separate lineages (e.g., there must be a common ancestor of chimps and humans).

Communal traits The tendency to act in a friendly and caring way.

Comparative psychology A branch of psychology that attempts to increase our understanding of ourselves by comparing other species to humans.

Computer model A hypothetical simulation of aspects of the human mind or behavior developed through a computer program.

Cooption In evolutionary biology, a term used to describe the situation where a trait that evolved for one function then evolves (is coopted) for a different purpose (sometimes referred to as an **exaptation** or a **preadaptation**).

Cultural relativism The notion that people's culture strongly influences their beliefs, perceptions and thoughts (is also used to describe the notion that we should avoid assuming that our own cultural practices are superior to those of other cultures).

Cultural universal Behavior or practices that are considered to occur in all cultures.

Darwinian medicine The field of study that concerns the relationship between evolution and human vulnerabilities to illness. Also known as **evolutionary medicine** (and in the case of mental health issues, **evolutionary psychiatry**).

Deoxyribonucleic acid (DNA) The macromolecule that contains all of our genetic information encoded in a series of base pairs.

Diagnostic and Statistical Manual of Mental Disorders (DSM-5) Manual categorizing all mental disorders (published by the American Psychiatric Association).

Dispersal The movement of individuals from their birth site to different breeding sites. Helps to avoid overcrowding and competition for resources with an organism's parents.

Dissociative identity disorder (also known as *multiple personality disorder*) A disorder where the sufferer considers there are two or more distinct personalities inhabiting his or her mind.

Dizygotic twins Twins born together but derived from separate eggs (i.e., non-identical twins).

Dual inheritance theory The notion that we "inherit" behavior via two routes: through our genes and from our culture.

Dunbar's number A proposed cognitive limit to the number of people a person can maintain a social relationship with. Generally regarded by Robin Dunbar to be around 150.

Embryogenesis The process of development of the embryo.

Empathizing Ability to understand and respond to the emotional state of others. Can be scored as an **empathizing quotient**.

Estrus The period of maximum female receptivity (that is, the period around which an ovum is released).

Extreme male brain hypothesis The notion that autistic individuals have an overly male-type brain that is therefore better able to systemize than empathize.

Fecundity The potential reproductive capacity of members of a given species.

Frequency dependent selection The evolutionary principle that suggests the effectiveness of a given phenotype is dependent upon the nature of other phenotypes within a population.

Gene A molecular unit of heredity consisting of a portion of DNA that codes for one polypeptide.

Gene-centered A perspective that regards the gene as the unit that natural selection acts on.

Gene–environment interaction In behavioral genetics, the relationship between genes and environmental input with regard to the development of a trait such as personality. Can be divided into **passive-gene–environment interaction**, whereby aspects of the environment (e.g., findings books in a home leads to reading) and **reactive-** or **evocative-gene–environment interaction**, whereby how others treat you affects your personality (e.g., being taught by a stimulating person).

Genetic determinism The notion that behavior is encoded in our genes and not subject to environmental influences.

Genotype All of the genes of an individual carried on his or her chromosomes.

Gaze plot A visual representation of how an individual scans an image.

Hamilton's rule Hamilton's rule is actually more correctly a formula that predicts when an animal is likely to behavior in a self-sacrificing way toward relatives. The formula is:

$$r \times B > C,$$

where B is the benefit gained by the recipient of the act, C is the cost to the actor and r is the proportion of genes shared between them by common descent. Hence, the closer the relationship in terms of genes shared, the more likely apparent altruism is to occur.

Heritability An estimation of the extent to which a trait is influenced by the genotype.

Heterozygous advantage A situation where a heterozygous genotype provides fitness advantages when compared with homozygous alternatives.

Hominid All existing and extinct great apes and modern humans.

Hominin Modern humans and all of our immediate ancestors, such as *Homo habilis*, *Homo erectus* and species of the genus *Australopithecus*.

Homozygous Having identical alleles (versions of a gene) at the same locus (position) on each of a pair of chromosomes.

Honest signal A trait or signal that is a genuine measure of quality.

Host–parasite arms race A situation where improvements on one side of a host and parasite relationship lead to selective pressure for counterimprovements on the other side. Leads to cycles of adaptation and counteradaptation.

Human Genome Project A multinational project to sequence the human genome (i.e., producing a sequence of all of the base pairs that our DNA contains). Determined that each human has around 20,500 genes.

Inbreeding depression Decline in vigor of offspring when closely related individuals breed together.

Inclusive fitness A measure of the proportion of an individual's genes that are passed on to future generations via offspring and other relatives.

In-group–out-group bias The tendency for humans to perceive members of their own group positively and members of other groups less so.

Intrasexual selection The part of sexual selection that refers to competition between members of one sex for access to members of the opposite sex. Most frequently leads to a lower threshold for aggression in males of a species.

Intersexual selection The part of sexual selection that refers to competition to attract members of the opposite sex. Commonly leads to "choosy" females and sexually ornamented males.

Isogamy A state where the two gametes are of the same size.

Kin detection mechanism An evolved neural mechanism that is used to estimate the degree of relatedness of another.

Kinship index A pairwise estimate of genetic relatedness between self and other. The outcome of the action of a kinship detection mechanism.

Language by-product hypothesis The notion that schizophrenia arises from having the wrong combination of genes that are related to language development.

Machiavellian intelligence hypothesis Theory that complex cognitive processes evolved in primates, including our own ancestral hominins, to deal with the challenges of social complexity rather than non-social challenges.

Major depressive disorder (also known as *unipolar depression*) Serious clinical depression where the sufferer has periods of very low mood that might be interspersed with relatively normal mood.

Males-compete/females-choose theory (MCFC) The idea that when it comes to mate choice decisions, males and females differ quite considerably such that we can consider males to be the competitive sex and females the choosy one.

Mating mind hypothesis Suggestion (by Geoffrey Miller) that language and intellect evolved through sexual selection with females, in particular, choosing males that were able to impress them through linguistic ability (which is taken as a measure of "fitness").

Market value An estimate of the value of a potential partner in terms of their ability to produce future offspring and provide for them. Generally considered to be based on characteristics such as physical attractiveness, resources and social standing.

Medicalizing of mental states The argument that "abnormal" behavior patterns are part of the normal range of human responses that have come to be labeled as medical conditions by psychiatrists.

Mendel's laws of inheritance Three principles of inheritance laid down by Gregor Mendel that became the foundation of the science of genetics. In a nutshell (modern-day interpretations of) the laws state that: chromosomes (and their genes) are separated into individual gametes prior to breeding; genes can either be dominant or recessive; genes are passed on independently of each other.

Microorganism An organism that is too small to be seen by the naked eye (e.g., viruses, bacteria, protozoa).

Mindreading/theory of mind (ToM) The ability of humans (and arguably members of some other species) to understand other individual's intentions, motives or emotions.

Mismatch hypothesis The notion that there is a mismatch between the conditions under which our species evolved and the environment we have created today (leading in many cases to physical and psychological health issues).

Mitochondrial Eve The most recent common female ancestor, related to all humans today. Lived in East Africa around 195,000 years ago.

Mixed-mating strategy The idea that some women, in supposedly monogamous relationships with a caring male, may increase the variability of their offspring by having extramarital affairs with other males.

Modern evolutionary synthesis The combination of Darwinism with genetics that explains evolution in terms of differential gene replication.

Monogamy A pair-bonded relationship where one male and one female mate exclusively with one another.

Monozygotic twins Twins derived from one fertilized egg (i.e., identical twins).

Moralistic fallacy The attempt to derive an empirical "is" argument from a moral "ought" argument. That is, if something appears to be morally good, then it must be factually correct.

Motile Technically "capable of self propelled action."

Multicellular An animal or plant that is made up of multiple cells.

Multilevel selection The notion that natural selection acts at a number of different levels, including the gene, individual, group and population.

Multiregion hypothesis (of human evolution) Theory that our species *Homo sapiens* evolved from *Homo erectus* simultaneously in a number of different geographical locations.

Mutual mate choice (MMC) Theory that humans are in the main a pair-bonded species where, due to relatively equal investment, the sexes are more similar than different.

Mutualism A situation where two or more individuals work together to achieve a goal that individuals would find difficult or impossible to accomplish on their own.

Natural selection Darwin's process to explain how evolution occurs. In modern terms can be considered as the differential survival of different phenotypes.

Naturalistic fallacy The attempt to derive a moral "ought" argument from an empirical "is" argument. That is, if something is factually correct, then we should use this to derive our moral code.

Neocortex Recently evolved part of the mammalian cortex; involved in higher functions such as reasoning and planning.

Neo-Darwinism Darwinian evolution that is informed by Mendelian genetics. (See also **modern evolutionary synthesis**.)

Neophobia Fear of the new (often applied in animal behavior to avoidance of novel foods).

Neoteny Retaining juvenile features into adulthood.

Nepotism Favoring kin over non-kin.

Nepotistic strategist The notion that we and other species may have evolved responses that help us to pass on our genes by favoring kin.

Neurocultural theory of emotional expression The theory (by Paul Ekman) that people of different cultures share a number of primary emotional expressions but how often we exhibit these varies from one culture to another.

Neuroimaging techniques Scanning techniques that allow experts to see either the structure or functioning of the human brain. Includes PET (positron emission tomography), CAT (computer axial tomography), MRI (magnetic resonance imaging), fMRI (functional magnetic resonance imaging) and MEG (magnetoencephalography).

Neuron (also spelled **neurone**) the technical name for a brain cell.

Neurotransmitter A chemical substance that allows neurons to communicate.

Out-of-Africa hypotheses (of human evolution) Theory that our species, *Homo sapiens,* evolved from *Homo erectus* in Africa before migrating to other parts of the globe.

Oviparity The situation where fertilization occurs inside the body of the mother and eggs are later laid.

Ovoviviparity The situation where eggs are brooded within the body.

Ovuliparity The situation where fertilization occurs outside the body of the mother (many fish and amphibians).

Parental investment Aid or resources provided to one offspring that might otherwise have been allocated to other offspring.

Parent–offspring conflict Predicted conflict over resource allocation where the offspring demands more than the parent is prepared to provide at that stage in development.

Parthenogenesis ("virgin birth") Asexual reproduction by production of offspring from unfertilized eggs.

Personality states The notion that behaviors that we see as enduring personality states are, in fact, largely is determined by a given situation.

Personality traits The notion that personality consists of relatively enduring ways of behaving and that personality can be encapsulated by a number of dimensions (such as extraversion and openness to experience).

Phenotype All of an organisms traits. The outcome of interaction between genotype and environment.

Pheromones Hormones that are designed to alter the internal state or behavior of other individuals (e.g., pheromones are considered to be involved in sexual attraction).

Planned deception A form of deception where the actor has, in some sense, planned to mislead (e.g., hiding as opposed to being camouflaged).

Polyandry The mating situation where one female has access to a number of males.

Polygyny The mating situation where one male has access to a number of females.

Polypeptide A small protein or part of a larger protein. Coded for by one gene.

Primary sexual characteristics Features that differ between the sexes that have evolved for reproductive purposes, such as the fallopian tubes or the testes.

Promiscuity The mating situation where there is no long-term commitment by either sex following mating, and mating can occur with multiple partners.

Proximate explanation An explanation for a behavior or internal state couched in terms of current biological, cognitive, social or developmental mechanisms. That is, a "here-and-now" explanation.

Pluralistic ignorance The finding that individuals frequently act the way they do, not because they believe it is the right way to behave but because they believe that everyone else believes it is the right way to behave.

Psychopathy A personality disorder where an individual acts in cold, unfeeling and exploitative ways.

Punnett square A diagram used to predict the outcome of a breeding experiment (named after Reginald Punnett).

Receptor site The position on a neuron where a neurotransmitter is received.

Reciprocity The situation where one individual aids another with no immediate return but rather receives aid at a later time (also called **reciprocal altruism**).

Red Queen hypothesis The proposal that organisms have to continue to evolve just to remain as successful as their ancestors were because other organisms (competitors, predators or parasites) are also evolving to gain an upper hand. The host–parasite arms race is a version of this.

Replicator An entity (such as a gene) that can make copies of itself.

Reproductive success How well an organism is able to pass on copies of its genes to the next generation.

Reproductive value Anticipated number of future offspring production as a function of age. First used by Ronald Fisher in 1930.

Runaway selection Hypothesis that male characteristics that originally arose via natural selection can later become highly elaborate due to female choice, propelling them onto extreme forms (proposed by Ronald Fisher).

Schizophrenia A serious mental health disorder where the sufferer exhibits a number of symptoms, such as delusions, hallucinations, disorganized incoherent thoughts and, in some cases, paranoia.

Secondary sexual characteristics Features that differ between the sexes that are not directly related to reproduction but rather evolved to help attract or gain access to the opposite sex. Frequently, these are exaggerated features in males, such as bright colored feathers, or weapons, such as large teeth or horns.

Selection pressure (also known as *selective pressure*) An apparent force that acts on organisms, driving them to evolve in a particular direction.

Sexual dimorphism A term introduced by Darwin to describe differences (physical and behavioral) between the sexes. Arose from sexual selection.

Sexual selection Darwin's second mechanism (or principle) of evolutionary change. Suggests that physical and psychological traits arose because they are either favored by members of the opposite sex or because they are beneficial when in competition with the same sex.

Signaling theory Proposal (by Amotz Zahavi) that suggests that sexual selection has endowed male animals with handicaps (such as long tails or other ornaments) that evolved to signal their underlying genetic quality to females. Also known as **honest signaling theory**, **costly signaling theory** and the **handicap hypothesis**.

Social comparison hypothesis A version of the mismatch hypothesis that suggests depression results from the unrealistic comparisons we make between ourselves and the glamorous images we see in the media.

Social brain hypothesis Notion that schizophrenia is due to having evolved a complex brain to deal with social complexity. In some cases, neurodevelopmental processes do not proceed properly leading to psychiatric problems.

Sociality The tendency of humans and members of many other species to associate with others and form into groups.

Social role theory The hypothesis that pressures from society lead us to take on the roles we do. In relation to gender roles, social role theorists generally reject biological/evolutionary explanations.

Superorganic The treatment of culture as if it were an organism in its own right, with its own autonomy and the ability to shape people.

Systemizing The ability or tendency to apply rule-governed skills to process information.

Trait Any observable feature from hair color and height to the tendency to act aggressively or kindly. In personality psychology, a trait is defined as a relatively stable element of personality.

Ultimate explanation Explaining behavior in relation to its adaptive value. The attempt to explain why the propensity for a particular form of behavior evolved in the first place.

Universality The notion that despite cultural difference, there are facets of human nature that are found in all cultures (and that such facets may then be seen as adaptations or as by-products of adaptations).

Variation Genetic diversity.

Viviparity The situation where offspring are born full term.

Withdrawal reflex A simple reflex action that involves a limited number of neurons, allowing an animal to withdraw a part of its body from danger very rapidly.

Word family The base form of a word plus all of its derived forms, such as different tenses.

Zygote The technical term for a fertilized egg.

References

Adelson, E. H. (2000) Lightness perception and lightness illusions. In M. Gazzaniga (Ed.), *The new cognitive neurosciences* (2nd ed., pp. 339–35). Cambridge, MA: MIT Press.

Ainsworth, M. D. S. (1967) *Infancy in Uganda: Infant care and the growth of love*. Baltimore: Johns Hopkins University Press.

Alexander, R. D. (1974) The evolution of social behaviour. *Annual Review of Ecology and Systematics, 5*, 325–83.

Allport, G. W., & Odbert, H. S. (1936) Trait-names: A psycho-lexical study. *Psychological Monographs, 47*(1), i.

Anderson, J. R., & Milson, R. (1989) Human memory: An adaptive perspective. *Psychological Review, 96*(4), 703.

Anderson, J. R., & Schooler, L. J. (1991) Reflections of the environment in memory. *Psychological Science, 2*(6), 396–408.

Andrade, M. C. B. (1996) Sexual selection for male sacrifice in redback spider. *Science, 271*, 70–72.

Andreasen, N. C. (1987) Creativity and mental illness: prevalence rates in writers and their first-degree relatives. *American Journal of Psychiatry, 144*, 1288–92.

Andreoni, J., & Miller, J. (1993) Rational cooperation in the finitely repeated prisoner's dilemma: Experimental evidence. *Econ. J., 103*, 570–85.

Andrew, R. J. (1963a) The origins and evolution of the calls and facial expressions of the primates. *Behaviour, 20*, 1–109.

Andrew, R. J. (1963b) Evolution of facial expressions. *Science, 142*, 1034–41.

Archer, J. (1996) Evolutionary social psychology. In M. Hewstone, S. Wolfgang & G. M. Stephenson (Eds.), *Introduction to social psychology* (pp. 24–45. Oxford: Blackwell.

Archer, J. (2004) Sex differences in aggression in real-world settings: A meta-analytic review. *Review of General Psychology, 8*, 291–322.

Archer, J. (2009) Does sexual selection explain human sex differences in aggression? *Behavioral and Brain Sciences, 32*, 249–311.

Ardrey R. (1970) *The social contract: A personal inquiry into the evolutionary sources of order and disorder*. Collins: London.

Armon-Jones, C. (1985) Prescription, explication and the social construction of emotion. *Journal for the Theory of Social Behaviour, 15*, 1–22.

Asch, S. E. (1951) Effects of group pressure upon the modification and distortion of judgments. In H. Guetzkow (Ed.), *Groups, Leadership, and Men* (pp. 177–90). Pittsburgh: Carnegie Press.

Auyeung, B., Baron-Cohen, S., Chapman, E., Knickmeyer, R., Taylor, K. & Hackett, G. (2009) Foetal testosterone and autistic traits. *British Journal of Psychology, 100,* 1–22.

Auyeung, B., Taylor, K., Hackett, G., & Baron-Cohen, S. (2010) Fetal testosterone and autistic traits in 18 to 24-month-old children. *Molecular Autism, 1,* 11.

Babiak, P., & Hare, R. D. (2006) *Snakes in suits: When psychopaths go to work.* New York: Regan Books/HarperCollins.

Baillargeon, R. (1987) Object permanence in 3½-and 4½-month-old infants. *Developmental Psychology, 23*(5), 655.

Bamberger, J. (1974) The myth of matriarchy: why men rule in primitive society. *Woman, Culture, Society,* 263–80.

Barkow, J. H., Cosmides, L. & Tooby, J. (Eds.). (1992) *The adapted mind: Evolutionary psychology and the generation of culture.* Oxford: Oxford University Press.

Baron-Cohen, S. (1995) *Mindblindness: An essay on autism and theory of mind.* Boston: MIT Press/Bradford Books.

Baron-Cohen, S. (2002) The extreme male brain theory of autism. *Trends in Cognitive Science, 6,* 248–254.

Baron-Cohen, S. (2003) *The essential difference: Men, women and the extreme male brain.* London: Penguin/Basic Books.

Baron-Cohen, S. (2012) Autism and the technical mind. *Scientific American, 307,* 72–7.

Baron-Cohen, S., Auyeung, B., Nørgaard-Pedersen, B., Hougaard. D. M., Abdallah, M. W., Melgaard, L., Cohen, A. S., Chakrabartim B., Ruta, L., & Lombardo, M. V. (2014) Elevated fetal steroidogenic activity in autism. *Molecular Psychiatry,* 1–8. Advance online publication 3 June 2014.

Baron-Cohen, S., & Wheelwright, S. (2004) The empathy quotient (EQ): An investigation of adults with Asperger syndrome or high functioning autism, and normal sex differences. *Journal of Autism and Developmental Disorders, 34,* 163–75.

Barrett, L. F. (2013) Psychological construction: The Darwinian approach to the science of emotion. Emotion Review, 5, 379–389.

Barrett, L., Dunbar, R. & Lycett, J. (2002) *Human evolutionary psychology.* New York: Palgrave.

Bateson, P., Mendl, M. & Feaver, J. (1990) Play in the domestic cat is enhanced by rationing of the mother during lactation. *Animal Behaviour, 40*(3), 514–25.

Bellis, M. A., & Baker, R. R. (1990) Do females promote sperm competition? Data for humans. *Animal Behaviour, 40*(5), 997–999.

Belsky, J., Houts, R. M. & Fearon, R. P. (2010) Infant attachment security and the timing of puberty testing an evolutionary hypothesis. *Psychological Science, 21*(9), 1195–1201.

Bentall, R. P. (2003) *Madness explained: Psychosis and human nature.* London: Penguin.

Bentall, R. P. (2009) *Doctoring the mind: Is our current treatment of mental illness really any good?* New York: New York Press.

Bereczkei, T. (1998) Kinship network, indirect childcare, and fertility among Hungarians and Gypsies. *Evolution and Human Behavior, 19,* 283–298.

Berkowitz, L. (2011) *Aggression: A social psychological analysis.* New York: McGraw-Hill.

Betzig, L. (2013) Fathers versus sons: Why Jocasta matters. In M. L. Fisher, J. R. Garcia, and R. S. Chang (Eds.), *Evolution's empress: Darwinian perspectives on the nature of women* (pp. 187–203). New York: Oxford University Press.

Block, N. (1996) How heritability misleads about race. *The Boston Review, 20*(6), 30–5.

Bloom, G., & Sherman, P. W. (2005) Dairying barriers affect the distribution of lactose malabsorption. *Evolution and Human Behavior, 26*(4), 301–312.

Boehm, C. (1999) *Hierarchy in the forest: The evolution of egalitarian behavior.* Cambridge, MA: Harvard University Press.

Bostwick, D. G., & Cheng, L. (2014) *Urological surgical pathology.* New York: Saunders.

Bowlby, J. (1969) *Attachment and loss* (3 vols.). London: Hogarth.

Boyd. R. & Richerson, P. J. (1983) Why is culture adaptive? *Quarterly Review of Biology, 58,* 209–214.

Boyd, R., Richerson, P. J. & Henrich, J. (2011) Rapid cultural adaptation can facilitate the evolution of large-scale cooperation. *Behavioral Ecology and Sociobiology, 65*(3), 431–444.

Broadbent, D. E. (1958) *Perception and communication.* New York: Pergamon Press.

Brown, D. E. (1991) *Human universals.* New York: McGraw-Hill.

Brown, N. R., & Sinclair, R. C. (1999) Estimating number of lifetime sexual partners: Men and women do it differently. *Journal of Sex Research, 36*(3), 292–7.

Brown, R., & Kulik, J. (1977) Flashbulb memories. *Cognition, 5*(1), 73–99.

Browne, K. R. (2009) Sex differences in aggression: Origins and implications for sexual integration of combat forces. *Behavioral and Brain Sciences, 32,* 270–71.

Brüne, M. (2004) Schizophrenia – An evolutionary enigma? *Neuroscience and Biobehavioral Reviews, 28,* 41–53.

Buller, D. J. (2005a) *Adapting minds: Evolutionary psychology and the persistent quest for human nature.* Cambridge, MA: MIT Press.

Buller, D. J. (2005b) Evolutionary psychology: The emperor's new paradigm. *Trends in Cognitive Science, 9,* 277–83.

Burns, J. (2007) *The descent of madness: Evolutionary origins of psychosis and the social brain.* New York: Routledge.

Burnstein, E., Crandall, C. & Kitayama, S. (1994) Some neo-Darwinian decision rules for altruism: Weighing cues for inclusive fitness as a function of the biological importance of the decision. *Journal of Personality and Social Psychology, 67,* 773–89.

Buss, D. M. (1989) Sex differences in human mate preferences: Evolutionary hypotheses tested in 37 cultures. *Behavioral and Brain Sciences, 12,* 1–49.

Buss, D. M. (2003) Sexual strategies: A journey into controversy. *Psychological Inquiry, 14,* 217–24.

Buss, D. M. (2004) *The evolution of desire* (2nd ed.). New York: Basic Books.

Buss, D. M. (2014) *Evolutionary psychology: The new science of the mind* (5th ed.). Boston: Allyn & Bacon.

Buss, D. M. (2013a) Feminist evolutionary psychology: Some reflections [Response to Sokol-Chang & Fisher]. *Journal of Social, Evolutionary, and Cultural Psychology, 7,* 295–6.

Buss, D. M. (2013b) The science of human mating strategies: An historical perspective. *Psychological Inquiry, 24,* 171–7.

Buss, D. M., & Hawley, P. (Eds.) (2011) *The evolution of personality and individual.* Oxford: Oxford University Press.

Buss, D. M., & Schmitt, D. P. (2011) Evolutionary psychology and feminism. *Sex Roles, 64,* 768–87.

Byrne, R., & Whitten, A. (1988) *Machiavellian intelligence.* Oxford: Oxford University Press.

Campbell, A. (2013a) *A mind of her own: The evolutionary psychology of women* (2nd ed.). Oxford: Oxford University Press.

Campbell, A. (2013b) Mutual mate choice: Sexual selection versus sexual conflict. *Psychological Inquiry, 24,* 178–82.

Carroll, L. (1871) *Through the looking glass and what Alice found there.* Macmillan: London.

Carter, R. T., & Qureshi, A. (1995) A typology of philosophical assumptions in multi-cultural counseling and training. In J. G. Ponterotto, J. M. Casas, L. A. Suzuki & C. M. Alexander (Eds.), *Handbook of multicultural counseling,* (pp. 239–62). Thousand Oaks, CA: Sage.

Catchpole, C. K., & Slater, P. J. B. (2008) *Bird song: Biological themes and variations.* Cambridge: Cambridge University Press.

Chakrabarti, B., Dudridge, F., Kent, L., Wheelwright, S., Hill-Cawthorne, G., Allison, C., Banerjee-Basu, S. & Baron-Cohen, S. (2009) Genes related to sex-steroids, neural growth and social-emotional behaviour are associated with autistic traits, empathy and Asperger Syndrome. *Autism Research, 2,* 157–77.

Chisholm, J. S. (1999) *Death, hope and sex: Steps to an evolutionary ecology of mind and morality.* Cambridge: Cambridge University Press.

Clutton-Brock, T. H. (1991) *The evolution of parental care.* Princeton: Princeton University Press.

Clutton-Brock, T. H. (2009) Cooperation between non-kin in animal societies. *Nature, 462,* 51–7.

Clutton-Brock, T. H., Gaynor, D., McIlrath, G. M., MacColl, A. D. C., Kansky, R., Chadwick, P., Manser, M., Brotherton, P. N. M. & Skinner, J. D. (1999) Predation, group size and mortality in a cooperative mongoose, *Suricata suricatta. Journal of Animal Ecology, 68,* 672–83.

Confer, J. C., Perilloux, C. & Buss, D. M. (2010) More than just a pretty face: Men's priority shifts toward bodily attractiveness in short-term versus long-term mating contexts. *Evolution and Human Behavior, 31,* 348.

Cooper, C. (2012) *Individual differences and personality* (3rd ed.). London: Hodder Arnold.

Cosmides, L. (1989) The logic of social exchange: Has natural selection shaped how humans reason? Studies with the Wason selection task. *Cognition, 31*(3), 187–276.

Cosmides, L., Barrett, H.C. & Tooby, J. (2010) Adaptive specializations, social exchange, and the evolution of human intelligence. *Proceedings of the National Academy of Sciences, 107*(Supplement 2), 9007–14.

Cosmides, L., & Tooby, J. (1992) Cognitive adaptations for social exchange. In J. Barkow, L. Cosmides & J. Tooby (Eds.). *The adapted mind* (pp. 163–228). New York: Oxford University Press.

Cosmides, L., Tooby, J., Fiddick, L. & Bryant, G. (2005) Detecting cheaters. *Trends in Cognitive Sciences, 9,* 505–6.

Cosmides, L., Tooby, J. & Kurzban, R. (2003) Perception of race. *Trends in Cognitive Sciences, 7,* 173–79.

Costa, P.T., & McCrae, R.R. (1985) *The NEO personality inventory: Manual, form S and form R.* Lutz, FL: Psychological Assessment Resources.

Costa, P.T., & McCrae, R.R. (1990) Personality disorders and the five-factor model of personality. *Journal of Personality Disorders, 4*(4), 362–71.

Crespi, B., Summers, K. & Dorus, S. (2007) Adaptive evolution of genes underlying schizophrenia. *Proceedings of the Royal Society*, B 2801–10.

Creswell, C., Shildrick, S. & Field, A. (2011) Interpretation of ambiguity in children: A prospective study of associations with anxiety and parental interpretations. *Journal of Child and Family Studies, 20,* 240–250.

Crooks, R.L., & Baur, K. (2013) *Our sexuality* (12th ed.). Belmont, CA: Wadsworth.

Daly, M., & Wilson, M. (1983) *Sex, evolution and behaviour* (2nd ed.). Belmont, CA: Wadsworth.

Daly, M., & Wilson, M. (1998) *The truth about Cinderella: A Darwinian view of parental love.* New Haven, CT: Yale University Press.

Daly, M., & Wilson, M. (2005) The "Cinderella effect" is no fairy tale. *Trends in Cognitive Sciences, 9,* 507–8.

Daly, M., & Wilson, M. (2007) Is the "Cinderella effect" controversial? In Crawford & Krebs (Eds.), *Foundations of evolutionary psychology* (pp. 383–400). Mahwah, NJ: Erlbaum.

Darwin, C. (1859) *On the origin of species by natural selection.* London: Murray.

Darwin, C. (1871) *The descent of man, and selection in relation to sex.* London: Murray.

Darwin, C. (1872) *The expression of the emotions in man and animals.* London: HarperCollins.

Dawkins, R. (2008) Why Darwin matters. *The Guardian,* Saturday, 9th Feb. http://www.theguardian.com/science/2008/feb/09/darwin.dawkins1 (Retrieved 28th May 2014.)

Dawkins, R. (1993) Viruses of the mind. In B. Dahlbom (Ed.), *Dennett and his critics.* Wiley-Blackwell.

Dawkins, R. (1976/2006) *The selfish gene.* Oxford: Oxford University Press.

Degler, C.N. (1991) *In search of human nature: The decline and revival of Darwinism in American social thought*. New York: Oxford University Press.

Diamond, J. (1997) *Guns, germs and steel: A short history of everybody for the last 13,000 years*. London: Vintage.

Dingemanse, N.J., Both, C., Drent, P.J. & Tinbergen, J.M. (2004) Fitness consequences of avian personalities in a fluctuating environment. *Proceedings of the Royal Society of London, Series B: Biological Sciences, 271*(1541), 847–52.

Dingemanse, N.J., Both, C., Drent, P.J., Van Oers, K. & Van Noordwijk, A.J. (2002) Repeatability and heritability of exploratory behaviour in great tits from the wild. *Animal Behaviour, 64*(6), 929–38.

Dominguez-Rodrigo, M. (2002) Hunting and scavenging by early humans: The state of the debate. *Journal of World Prehistory, 16,* 1–54.

Dudley, R. (2002) Fermenting fruit and the historical ecology of ethanol ingestion: Is alcoholism in modern humans an evolutionary hangover? *Addiction, 97,* 381–388.

Dunbar, R.I.M. (1993) The co-evolution of neocortical size, group size and language in humans. *Behavioral Brain Sciences, 16,* 681–735.

Dunbar, R. I. M., Barrett, L. & Lycett, J. (2005) *Evolutionary psychology: A beginner's guide.* Oxford: One World Books.

Dunbar, R.I.M., Clark, A. & Hurst, N.L. (1995) Conflict and cooperation among the Vikings: Contingent behavioural decisions. *Ethology and Sociobiology, 16,* 233–46.

Dunson, D. B., Colombo, B., & Baird, D. D. (2002) Changes with age in the level and duration of fertility in the menstrual cycle. *Human Reproduction, 17*(5), 1399–1403.

Eagly, A.H. (1987) *Sex differences in social behavior: A social-role interpretation*. Hillsdale, NJ: Erlbaum.

Eagly, A.H., & Wood, W. (1991) Explaining sex differences in social behavior: A meta-analytic perspective. *Personality and Social Psychology Bulletin, 17,* 306–15.

Eagly, A.H., & Wood, W. (1999) The origins of sex differences in human behavior: Evolved dispositions versus social roles. *American Psychologist, 54,* 408–23.

Eagly, A.H., & Wood, W. (2009) Sexual selection does not provide an adequate theory of sex differences in aggression. *Behavioral and Brain Sciences, 32,* 276–77.

Eagly, A.H., & Wood, W. (2011) Feminism and the evolution of sex differences and similarities. *Sex Roles, 64,* 758–67.

Eagly, A.H., & Wood, W. (2012) Social role theory. In P. van Lange, A. Kruglanski & E.T. Higgins (Eds.), *Handbook of theories in social psychology* (pp. 458–76). Thousand Oaks, CA: Sage.

Eagly, A.H., & Wood, W. (2013) The nature–nurture debates: 25 years of challenges in understanding the psychology of gender. *Perspectives on Psychological Science, 8,* 340–357.

Eastwick, P.W. (2013) The psychology of the pair-bond: Past and future contributions of close relationships research to evolutionary psychology. *Psychological Inquiry, 24,* 183–91.

Eibl-Eibesfeldt, I. (1973) *Social communication and movement*. New York: Academic Press.

Eibl-Eibesfeldt, I. (1989) *Human ethology*. New York: Aldine de Gruyter.

Eichhammer, P., Sand, P. G., Stoertebecker, P., Langguth, B., Zowe, M. & Hajak, G. (2005) Variation at the DRD4 promoter modulates extraversion in Caucasians. *Molecular Psychiatry, 10,* 520–22.

Ekman, P. (1972) Universals and cultural differences in facial expressions of emotion. In J. Cole (Ed.), *Nebraska symposium on motivation* (Vol. 19, pp. 207–82). Lincoln: University of Nebraska Press.

Ekman, P. (1992) An argument for basic emotions. *Cognition and Emotion, 6,* 169–200.

Ekman, P. (1999) Basic emotions. In T. Dalgleish & M. Power (Eds.), *Handbook of cognition and emotion, facial expressions* (pp. 45–60). New York: Wiley.

Ekman, P., & Friesen, W. V. (1971) Constants across cultures in the face and emotion. *Journal of Personality and Social Psychology, 17,* 124–9.

Ekman, P., & Friesen, W. V. (1986) A new pan-cultural facial expression of emotion. *Motivation and Emotion, 10,* 159–68.

Emlen, S. T., & Wrege, P. H. (2004) Division of labour in parental care behaviour of a sex-role-reversed shorebird, the wattled jacana. *Animal Behaviour, 68,* 847–55.

Emlen, S. T., & Wrege, P. H. (2004) Size dimorphism, intrasexual competition, and sexual selection in wattled jacana (*Jacana jacana*), a sex-role-reversed shorebird in Panama. *The Auk, 121*(2), 391–403.

Errington, F., & Gewertz, D. (1985) The chief of the Chambri: Social change and cultural permeability among a New Guinea people. *American Ethnologist, 12*(3), 442–54.

Eysenck, H. J. (1990) Genetic and environmental contributions to individual differences: The three major dimensions of personality. *Journal of Personality, 58*(1), 245–261.

Eysenck, H. J. (1991) Dimensions of personality: 16, 5 or 3?—Criteria for a taxonomic paradigm. *Personality and Individual Differences, 12*(8), 773–790.

Fagen, R. M. (1977) Selection for optimal age-dependent schedules of play behavior. *American Naturalist,* 395–414.

Fehr, E., & Gächter, S. (2000) Fairness and retaliation: The economics of reciprocity. *Journal of Economic Perspectives, 14,* 159–81.

Fidler, M., Light, P. & Costall, A. (1996) Describing dog behavior psychologically: Pet owners versus non-owners. *Anthrozoos: A Multidisciplinary Journal of The Interactions of People & Animals, 9,* 196–200.

Field, Z. C. & Field, A. P. (2013) How trait anxiety, interpretation bias and memory affect acquired fear in children learning about animals. *Emotion, 13,* 409–23.

Figueredo, A. J., Gladden, P. R. & Hohman. (2011) The evolutionary psychology of criminal behaviour. In S. C. Roberts, (Ed.), *Applied evolutionary psychology*. Oxford: Oxford University Press.

Fisher, R. A. (1930) *The genetical theory of natural selection*. Oxford: Clarendon Press.

Fitch, W. T., Hauser, M. D. & Chomsky. N. (2005) The evolution of the language faculty: Clarifications and implications. *Cognition, 97,* 179–210.

Flint, J., Greenspan, R. J. & Kendler, K. S. (2010) *How genes influence behavior*. Oxford: Oxford University Press.

Freeman, D. (1983) *Margaret Mead and Samoa: The making and unmaking of an anthropological myth*. Cambridge, MA: Harvard University Press.

Fredrickson, B. L. (1998) What good are positive emotions? *Review of General Psychology, 2*, 300–19.

Fredrickson, B. L. (2006) Unpacking positive emotions: Investigating the seeds of human flourishing. *Journal of Positive Psychology, 1*, 57–60.

Fredrickson, B. L. (2013) Positive emotions broaden and build. In E. Ashby Plant & P. G. Devine (Eds.), *Advances on experimental social psychology* (Vol. 47, pp. 1–53). Burlington: Academic Press.

Gächter, S., & Falk, A. (2002) Reputation and reciprocity: Consequences for the labour relation. *Scandinavian Journal of Economics, 104*, 1–26.

Gamble, C. (1999) *The Palaeolithic societies of Europe*. Cambridge: Cambridge University Press.

Gardner, H. (2010) A debate on "multiple intelligences." In J. Traub (Ed.), *Cerebrum: Forging ideas in brain science* (pp. 34–61). Washington: Dana Press.

Gaulin, S. J., McBurney, D. H. & Brakeman-Wartell, S. L. (1997) Matrilateral biases in the investment of aunts and uncles. *Human Nature, 8*(2), 139–151.

Geary, D. C. (1998) *Male, female*. Washington, DC: American Psychological Association.

Geissmann, T. (1993) *Evolution of communication in gibbons (Hylobatidae)*. (Unpublished doctoral dissertation). Zurich, University, Zurich, Switzerland.

Gershon, E. S., Martinez, M., Goldin, L. R. & Gejman, P. V. (1990) Genetic mapping of common diseases: The challenges of manic-depressive illness and schizophrenia. *Trends Genet, 6*, 282–87.

Gluckman, P., & Hanson, M. (2008) *Mismatch: The lifestyle diseases timebomb*. Oxford: Oxford University Press.

Gopnik, A., Meltzoff, A. N. & Kuhl, P. K. (2001) *How babies think: The science of childhood*. London: Phoenix.

Gould, J. L., & Gould, G. C. (1997) *Sexual selection: Mate choice and courtship in nature*. New York: W. H. Freeman.

Gray, J., (1992) *Men are from Mars, women are from Venus*. New York: HarperCollins.

Greenless, I. A., & McGrew, W. C. (1994) Sex and age differences in preferences and tactics of mate attraction: Analysis of published advertisements. *Ethology and Sociobiology, 15*, 59–72.

Griggs, R. A., & Cox, J. R. (1982) The elusive thematic-materials effect in Wason's selection task. *British Journal of Psychology, 73*(3), 407–20.

Haeckel, E. (1904) *Die Lebenswunder*. Stuttgart: Alfred Kröner.

Haidt, J., McCauley, C. & Rozin, P. (1994) Individual differences in sensitivity to disgust: A scale sampling seven domains of disgust elicitors. *Personality and Individual Differences, 16*(5), 701–13.

Hamilton, W. D. (1964). The genetical evolution of social behaviour. *Journal of Theoretical Biology, 7*, 1–52.

Hamilton, W. D., & Zuk, M. (1982) Heritable true fitness and bright birds: A role for parasites? *Science, 218*, 384–7.

Haney, C., Banks, W.C. & Zimbardo, P.G. (1973) Study of prisoners and guards in a simulated prison. *Naval Research Reviews, 9,* 1–17.

Hare, R.D.(1993) *Without conscience: The disturbing world of the psychopaths among us.* New York: Pocket Books.

Hare, R.D. (2003) *Manual for the Revised Psychopathy Checklist* (2nd ed.). Toronto, ON, Canada: Multi-Health Systems.

Harrington, J.M. (2001) Health effects of shift work and extended hours of work. *Occupational and Environmental Medicine, 58,* 58–72.

Harris, J.R. (2009) *The nurture assumption: Why children turn out the way they do.* New York: Simon & Schuster.

Hausfater, G., & Hrdy, S. (Eds.) (1984) *Infanticide: Comparative and evolutionary perspectives.* New York: Aldine.

Hebb, D.O. (1946) Emotion in man and animals: an analysis of the intuitive processes of recognition. *Psychological Review, 53,* 88–106.

Hess, U., & Thibault, P. (2009) Darwin and emotion expression. *American Psychologist, 642,* 120–128.

Hewstone, M., Stroebe, W. & Jonas, K. (2012) *Introduction to social psychology.* Chichester: Wiley.

Hidaka, B.H. (2012) Depression as a disease of modernity: explanations for increasing prevalence. *Journal of Affective Disorders, 140,* 205–14.

Hill, K. (2002) Altruistic cooperation during foraging by the Ache, and the evolved human predisposition to cooperate. *Human Nature, 13,* 105–28.

Hoebel, E.A. (1966) *Anthropology: The study of man.* New York: McGraw-Hill.

Horan, R.D., Bulte, E. & Shogren, J.F. (2005) How trade saved humanity from biological exclusion: An economic theory of Neanderthal extinction. *Journal of Economic Behavior & Organization, 58*(1), 1–29.

Humphrey, N. (1976) The social function of intellect. In P.P.G. Bateson & R.A. Hinde (Eds.), *Growing points in ethology* (pp. 303–17). Cambridge: Cambridge University Press.

Hyde, J.S. (2005) The gender similarities hypothesis. *American Psychologist, 60*(6), 581.

Ilardi, S.S. (2010) *The Depression Cure: The 6-step program to beat depression without drugs.* Cambridge, MA: Da Capo Press.

Ilardi, S.S., Jacobson, J.D., Lehman, K.A., Stites, B.A., Karwoski, L., Stroupe, N.N. & Young, C. (2007) *Therapeutic lifestyle change for depression: Results from a randomized controlled trial.* Paper presented at the annual meeting of the Association for Behavioral and Cognitive Therapy, Philadelphia, PA.

Ivey, P. (2000) Cooperative reproduction in Ituri Forest hunter-gatherers: Who cares for Efe infants? *Current Anthropology, 41,* 856–66.

Jamison, K.R. (1993) *Touched with fire: Manic-depressive illness and the artistic temperament.* New York: The Free Press.

Jamison, K.R. (1995) Manic-depressive illness and creativity. *Scientific American, 272,* 62–7.

Jamison, K.R. (2011) Great wits and madness: More near allied? *British Journal of Psychiatry, 199,* 351–2.

Jemal, A., Bray, F., Center, M.M., Ferlay, J., Ward, E. & Forman, D. (2011) Global cancer statistics. *CA: A Cancer Journal for Clinicians, 61,* 69–90.

Jones, A.C., & Gosling, S.D. (2005) Temperament and personality in dogs (Canis familiaris): A review and evaluation of past research. *Applied Animal Behaviour Science, 95,* 1–53.

Jones, D., & Hill, K. (1993) Criteria of facial attractiveness in five populations. *Human Nature, 4,* 271–96.

Kaessmann, H., Wiebe, V., Weiss, G. & Pääbo, S. (2001) Great ape DNA sequences reveal a reduced diversity and an expansion in humans. *Nature Genetics, 27*(2), 155–56.

Kandel, E. R. (1976) *Cellular basis of behaviour.* San Francisco: Freeman.

Kano, F., & Tomonaga, M. (2009) Face scanning in chimpanzees and humans: Continuity and discontinuity. *Animal Behavior, 79,* 227–35.

Kawai, M. (1965) Newly-acquired pre-cultural behavior of the natural troop of Japanese monkeys on Koshima Islet. *Primates, 6*(1), 1–30.

Kellogg, W.N., & Kellogg, L.A. (1933) *The ape and the child: A study of environmental influence upon early behavior.* New York: McGraw-Hill.

Kenrick, D.T. (2013) Men and women are only as different as they look! *Psychological Inquiry, 24,* 202–6.

Kenrick, D.T., Gabrielidis, C., Keefe, R. C. & Cornelius, J. (1996) Adolescents' age preferences for dating partners: Support for an evolutionary model of life-history strategies. *Child Development, 67,* 1499–1511.

Kenrick, D.T., & Keefe, R.C. (1992) Age preferences in mates reflect sex differences in mating strategies. *Behavioral & Brain Sciences, 15,* 75–91.

King, R., (2004) Machiavelli: Philosopher of power. New York: HarperCollins.

Klatsky, A.L. (2003) Drink to your health? *Scientific American, 288,* 74–81.

Kogan, A., Saslow, L.R., Impett, E.A., Oveis, C., Keltner, D. & Saturn, S.R. (2011) Thin-slicing study of the oxytocin receptor (OXTR) gene and the evaluation and expression of the prosocial disposition. *Proceedings of the National Academy of Sciences, 108,* 19189.

Krahe, B. (2001) *The social psychology of aggression.* Hove: Psychology Press.

Kring, A. M., Johnson, S. L. Davison, G. C. & Neale, J. M. (2014). *Abnormal psychology: DSM-5 update* (12th ed.). New York: Wiley.

Kuhl, P. K., Williams, K. A., Lacerda, F., Stevens, K. N. & Lindblom, B. (1992) Linguistic experience alters phonetic perception in infants by 6 months of age. *Science, 255*(5044), 606–608.

Kurzban, R., Tooby, J. & Cosmides, L (2001) Can race be erased? Coalitional computation and social categorization. *Proceedings of the national Academy of Sciences, 89,* 15387–92.

Lack, D. (1968) *Ecological adaptations for breeding in birds.* London, UK: Methuen.

Lai, M., Lombardo, M., Chakrabarti, B., Ecker, C., Sadek, S., Wheelwright, S., Murphy, D., Suckling, J., Bullmore, E., MRC AIMS Consortium, & Baron-Cohen, S. (2012) Individual differences in brain structure underpin empathizing-systemizing cognitive styles in male adults. *NeuroImage, 61,* 1347–54.

Laland, K.N., & Brown, G.R. (2011) *Sense and nonsense: Evolutionary perspectives on human behaviour.* Oxford: Oxford University Press.

Larsson, H., Andershed, H. & Lichtenstein, P. (2006) A genetic factor explains most of the variation in the psychopathic personality. *Journal of Abnormal Psychology, 115*(2), 221.

Lazarus, R.S. (1991) *Emotion and adaptation.* New York: Oxford University Press.

Lee, R.B. (1979) *The !Kung San. Men, women and work in foraging society.* Cambridge: Cambridge University Press.

Lewontin, R. (1970) Race and intelligence. *Bulletin of the Atomic Scientists, 26*(3), 2–8.

Lickliter, R., & Honeycutt, H. (2003) Developmental dynamics: Toward a biologically plausible evolutionary psychology. *Psychological Bulletin, 129,* 819–835.

Lieberman, D., Tooby, J. & Cosmides, L. (2007) The architecture of human kin detection. *Nature, 393,* 639–640.

Liesen, L.T. (2007) Women, behavior, and evolution: Understanding the debate between feminist evolutionists and evolutionary psychologists. *Politics and the Life Sciences, 26,* 51–70.

Lippa, R.A. (2009) Sex differences in sex drive, sociosexuality, and height across 53 nations: Testing evolutionary and social structural theories. *Archives of Sexual Behavior, 38*(5), 631–51.

Lombardo, M., Ashwin, E., Auyeung, B., Chakrabarti, Taylor, K., Hackett, G., Bullmore, E. & Baron-Cohen, S. (2012) Fetal testosterone influences sexually dimorphic gray matter in the human brain. *Journal of Neuroscience, 32,* 674–80.

Machiavelli, N. (1988). The prince. Q. Skinner & R. Price (Eds.). Cambridge: Cambridge University Press.

Marr, D. (1982) *Vision: A computational investigation into the human representation and processing of visual information.* New York: Henry Holt.

Maynard-Smith, J. (1964) Group selection and kin selection. *Nature, 201,* 1145–47.

Maynard Smith, J. (1998) The origin of altruism. *Nature, 393,* 639–40.

Mead, M. (1935) *Sex and temperament.* London: Routledge and Kegan Paul.

Mealey, L. (1995) The sociobiology of sociopathy: An integrated evolutionary model. *Behavioral and Brain Sciences, 18*(03), 523–41.

Milan, E.L. (2010) *Looking for a few good Males: Female Choice in evolutionary biology.* Baltimore: Johns Hopkins University Press.

Milgram, S. (1963) Behavioral study of obedience. *The Journal of Abnormal and Social Psychology, 67*(4), 371.

Miller, G.F. (2000) *The mating mind: How sexual choice shaped the evolution of human nature.* London: Heinemann/Doubleday.

Miller, G.F. (2013) Mutual mate choice models as the red pill in evolutionary psychology: Long delayed, much needed, ideologically challenging, and hard to swallow. *Psychological Inquiry, 24,* 207–10.

Min, K.J., Lee, C.K. & Park, H.N. (2012) The lifespan of Korean eunuchs. *Current Biology, 22,* 792–3.

Mischel, W. (1968) *Personality and assessment.* New York: Wiley.

Monteleone, M. R., Clark, T. G., Moore, L. R., Payne, E., Walton, R. & Flint, J. (2006) Genetic polymorphisms and personality in healthy adults: A systematic review and meta-analysis. *Molecular Psychiatry, 8,* 471–84.

Moon, C., Lagercrantz, H. & Kuhl, P. K. (2013) Language experienced in utero affects vowel perception after birth: A two-country study. *Acta Paediatrica, 102*(2), 156–60.

Morton, J., & Johnson, M. H. (1991) CONSPEC and CONLERN: A two-process theory of infant face recognition. *Psychological Review, 98*(2), 164.

Nagell, K., Olguin, R. S. & Tomasello, M. (1993) Processes of social learning in the tool use of chimpanzees – Pan troglodytes – and human children – Homo sapiens. *Journal of Comparative Psychology, 107*(2), 174.

Nairne, J. S., & Pandeirada, J. N. (2008) Adaptive memory: Is survival processing special? *Journal of Memory and Language, 59*(3), 377–85.

Nash, A., & Grossi, G. (2007) Picking Barbie's brain: Inherent sex differences in scientific ability? *Journal of Interdisciplinary Feminist Thought, 2,* 5.

Nesse, R. M. (2005) Natural selection and the regulation of defenses: A signal detection analysis of the smoke detector principle. *Evolution and Human Behavior, 26,* 88–105.

Nesse, R. M. (2006) Darwinian medicine and mental disorders. *International Congress Series, 1296,* 83–94.

Nesse, R. M. (2009) Evolutionary origins and functions of emotions. In D. Sander & K. R. Scherer (Eds.), *The Oxford companion to emotion and the affective sciences* (pp. 158–64). Oxford: Oxford University Press.

Nesse, R. M. (2011) Why has natural selection left us so vulnerable to anxiety and mood disorders? *Canadian Journal of Psychiatry, 56,* 705–6.

Nesse, R. M. (2012) Evolution: A basic science for medicine. In A. Poiani (Ed.). *Pragmatic evolution: Applications of evolutionary theory* (107–14). Cambridge: Cambridge University Press.

Nesse, R. M. (2014) A general "Theory of Emotion" is neither necessary nor possible. *Emotion Review, 6*(4) 320–22.

Nesse, R. M., & Dawkins, R. (2010) Evolution: Medicine's most basic science. In D. A. Warrell, T. M. Cox, J. D. Firth & E. J. J. Benz (Eds.), *Oxford textbook of medicine* (5th ed., pp. 12–15). Oxford: Oxford University Press.

Nesse, R. M., & Williams, G. C. (1995) *Evolution and healing: The new science of Darwinian medicine.* London: Weidenfeld & Nicolson.

Nettle, D. (2005) An evolutionary perspective on the extraversion continuum. *Evolution and Human Behavior, 26,* 363–73.

Nettle, D. (2009) *Evolution and genetics for psychology.* Oxford: Oxford University Press.

Nettle, D., Coall, D. A. & Dickins, T. E. (2011) Early-life conditions and age at first pregnancy in British women. *Proceedings of the Royal Society, B 278,* 1721–7.

Neuberg, S. L., Kenrick, D. T. & Schaller, M. (2010) *Evolutionary social psychology.* In S. T. Fiske, D. Gilbert & G. Lindzey (Eds.), *Handbook of social psychology* (5th ed., pp. 761–96). New York: Wiley.

Newell, A., Shaw, J. C. & Simon, H. A. (1959, January) Report on a general problem-solving program. In *IFIP Congress,* 256–64.

Oda, R. (2001) Sexually dimorphic mate preference in Japan: An analysis of lonely hearts advertisements. *Human Nature, 12,* 191–206.

O'Donovan, M.M. (2013) Feminism, disability, and evolutionary psychology: What's missing? *Disability Studies Quarterly, 33,* 4.

Ohman, A., & Mineka, S. (2001) Fears, phobias and preparedness: Toward an evolved module of fear and fear learning. *Psychological Review, 108,* 483–522.

Okasha, S. (2002) Genetic relatedness and the evolution of altruism. *Philosophy of Science, 69,* 138–49.

O'Steen, S., Cullum, A.J., & Bennett, A.F. (2002) Rapid evolution of escape ability in Trinidadian guppies (*Poecilia reticulata*). *Evolution, 56*(4), 776–84.

Paley, W. (1802) *Natural Theology: Or, evidences of the existence and attributes of the deity: Collected from the appearances of nature.* London: Gould and Lincoln.

Panksepp, J. (1998) *Affective neuroscience: The foundations of human and animal emotions.* New York: Oxford University Press.

Panksepp, J. (Ed.) (2004) Frontmatter. In *Textbook of biological psychiatry.* Hoboken, NJ: Wiley.

Panksepp, J. (2005) Affective consciousness: Core emotional feelings in animals and humans. *Consciousness & Cognition, 14,* 19–69.

Panksepp, J., & Biven, L. (2010) *The archaeology of mind: Neuroevolutionary origins of human emotion.* New York: W.W. Norton.

Pawlowski, B., & Dunbar, R.I.M. (1999) Impact of market value on human mate choice decisions. *Proceedings of the Royal Society of London, B, 266,* 281–5.

Pearce, J.M. (2008) *Animal learning and cognition: An introduction* (3rd ed.). Hove: Psychology Press.

Perrett, D.I., Lee, K.J., Penton-Voak, I.S., Rowland, D.R., Yoshikawa, S., Burt, D.M., Henzi, S.P., Castles, D.L. & Akamatsu, S. (1998) Effects of sexual dimorphism on facial attractiveness. *Nature, 394,* 884–7.

Pinker, S. (1997) *How the mind works.* New York: Norton.

Pinker, S. (2002) *The blank slate: The denial of human nature in modern intellectual life.* London: Viking Press.

Pinker, S. (2011) *The better angels of our nature: The decline of violence in history and its causes.* London: Penguin.

Pinker, S. (2012) *The false allure of group selection. Edge,* June 19th. http://edge.org/conversation/the-false-allure-of-group-selection.

Plomin, R. (2011) Commentary: Why are children in the same family so different? Non-shared environment three decades later. *International Journal of Epidemiology, 40*(3), 582–92.

Premo, L.S., & Hublin, J.J. (2009) Culture, population structure, and low genetic diversity in Pleistocene hominins. *Proceedings of the National Academy of Sciences, 106*(1), 33–37.

Read, L.E. (1958) *I, pencil.* Irvington-on-Hudson, NY: Foundation for Economic Education.

Richards, R.J. (1986) A defense of evolutionary ethics. *Biology and Philosophy, 1*(3), 265–93.

Richerson, P.J., & Boyd, R. (2001) Built for speed, not for comfort: Darwinian theory and human culture. *History and Philosophy of the Life Sciences,* 425–465.

Richerson, P. J., & Boyd, R. (2008) *Not by genes alone: How culture transformed human evolution*. Chicago: University of Chicago Press.

Ridley, M. (1993) *The Red Queen: Sex and the evolution of human nature*. London: Penguin.

Ridley, M. (1996) *The origins of virtue*. London: Viking Press.

Ridley, M. (2010) *When ideas have sex*. TedGlobal [video file] retrieved from http://www.ted.com/talks/matt_ridley_when_ideas_have_sex (May, 2014.)

Roberts, S. C. (Ed.) (2011) *Applied evolutionary psychology*. Oxford: Oxford University Press.

Roberts, S. C., & Havlíček, J. (2013) Humans are Dunnocks not Peacocks: On cause and consequence of variation in human mating strategies. *Psychological Inquiry, 24,* 231–36.

Robinson, G., Fernald, R. & Clayton, D. (2008) Genes and social behavior. *Science, 322* (5903), 896–900.

Rosenberg, J., & Tunney, R. J. (2008) Human vocabulary use as display. *Evolutionary Psychology, 6,* 538–549.

Ross, L., & Nisbett, R. E. (1991) *The person and the situation: Perspectives of social psychology*. New York: Mcgraw-Hill.

Rossano, M. (2003) *Evolutionary psychology: The science of human behavior and evolution*. New York: Wiley.

Ruigrok, A., Salimi-Khorshidi, G., Lai, M-C., Baron-Cohen, S., Lombardo, M., Tait, R. & Suckling, J. (2014) A meta-analysis of sex differences in human brain structure. *Neuroscience and Biobehavioral Reviews, 39,* 34–50.

Ruta, L., Ingudomnukul, E., Taylor, E., Chakrabarti, B. & Baron-Cohen, S. (2011) Increased serum androstenedione in adults with autism spectrum conditions. *Psychoneuroendocrinology, 36,* 1154–63.

Salganik, M. J., Dodds, P. S. & Watts, D. J. (2006) Experimental study of inequality and unpredictability in an artificial cultural market. *Science, 311*(5762), 854–56.

Schacter, D. L. (2001) *The seven sins of memory*. New York: Houghton Mifflin.

Schaller, M., Simpson, J. A. & Kenrick, D. T. (2006) *Evolution and social psychology (frontiers of social psychology)*. New York: Psychology Press.

Shariff, A. F., & Tracy, J. L. (2011) What are emotion expressions for? *Current Directions in Psychological Science, 20,* 395–99.

Silk, J. B. (1980) Adoption and kinship in Oceania. *American Anthropologist, 82,* 799–820.

Silk, J. B. (1990) Human adoption in evolutionary perspective. *Human Nature, 1,* 25–52.

Singh, D. (1993a) Adaptive significance of female physical attractiveness: Role of waist-to-hip ratio. *Journal of Personality and Social Psychology, 65,* 293–307.

Singh, D. (1993b) Body shape and women's attractiveness: The critical role of waist-to-hip ratio. *Human Nature, 4,* 297–321.

Singh, D., & Singh, D. (2011) Shape and significance of feminine beauty: An evolutionary perspective. *Sex Roles, 64,* 723–31.

Smith, L., & Thelen, E. (2003) Development as a dynamic system. *Trends in Cognitive. Science, 7,* 343–48.

Sober, E., & Wilson, D. S. (1999) *Unto others: The evolution and psychology of unselfish behavior.* Cambridge, MA: Harvard University Press.

Somers, J. M., Goldner, E. M., Waraich, P. & Hsu, L. (2006) Prevalence and incidence studies of anxiety disorders: a systematic review of the literature. *Canadian Journal of Psychiatry 51,* 100–113.

Spelke, E. S., Breinlinger, K., Macomber, J. & Jacobson, K. (1992) Origins of knowledge. *Psychological Review, 99*(4), 605.

Stewart-Williams, S., & Thomas, A. G. (2013a) The ape that thought it was a peacock: Does evolutionary psychology exaggerate human sex differences? *Psychological Inquiry, 24,* 137–68.

Stewart-Williams, S., & Thomas, A. G. (2013b) The ape that kicked the hornet's nest: Response to commentaries on 'The ape that thought it was a peacock'. *Psychological Inquiry, 24,* 248–71.

Storey, S., & Workman, L. (2013) The effects of temperature priming on cooperation in the Iterated Prisoner's Dilemma. *Evolutionary Psychology,* 11, 52–67.

Strier, K. B. (2011) *Primate behavioral ecology* (4th ed.). Boston: Pearson.

Stringer, C. (2012) *The origin of our species.* London: Pearson.

Swami, V., & Salem, N. (2011) The evolutionary psychology of human beauty. In V. Swami (Ed.), *Evolutionary psychology: A critical introduction* (pp. 131–82). Chichester: Wiley-Blackwell.

Tajfel, H. (1970) Experiments in intergroup discrimination. *Scientific American, 223,* 96–102.

Tajfel, H. (1982) *Social identity and intergroup relations.* Cambridge: Cambridge University Press.

Tanner, N., M. (1981) *On becoming human.* Cambridge: Cambridge University Press.

Taylor, S. R. (2015) *Forensic psychology: The basics.* London: Routledge.

Toates, F. (2011) *Biological psychology: An integrative approach* (3rd ed.). Harlow: Pearson Education.

Toates, F. (2014) *How sexual desire works: The enigmatic urge.* Cambridge: Cambridge University Press.

Tomasello, M. (1999) *The cultural origins of human cognition.* Cambridge, MA: Harvard University Press.

Tooby, J. & DeVore, I. (1987) The reconstruction of hominid behavioural evolution through strategic modelling. In W. G. Kinzey (Ed.), *The evolution of human behaviour: Primate models* (pp. 183–237). Albany: State University of New York Press.

Trivers, R. L. (1971) The evolution of reciprocal altruism. *Quarterly Review of Biology, 46,* 35–57.

Trivers, R. L. (1972) Parental investment and sexual selection. In B. Campbell (Ed.) *Sexual selection and the descent of man, 1871–1971* (pp. 136–79). Chicago: Aldine-Atherton.

Trivers, R. L. (1974) Parent-offspring conflict. *American Zoologist,14,* 247–62.

Trivers, R. L. (1985) *Social evolution.* Menlo Park, CA: Benjamin/Cummings.

Turkheimer, E. (2000) Three laws of behavior genetics and what they mean. *Current Directions in Psychological Science, 9*(5), 160–64.

Turkheimer, E., Haley, A., D'Onofrio, B., Waldron, M. & Gottesman, I. I. (2003) Socioeconomic status modifies heritability of IQ in young children. *Psychological Science, 14*, 623–28.

Turner, D. C., & Bateson, P. P. G. (2000) *The domestic cat: The biology of its behaviour.* Cambridge: Cambridge University Press.

Van Hooff, J.A.R.A.M. (1967) Facial displays of catarrhine monkeys and apes. In D. Morris (Ed.), *Primate ethology* (pp. 7–68). London: Weidenfeld & Nicolson.

Van Hooff, J.A.R.A.M. (1972) A comparative approach to the phylogeny of laughter and smile. In R.A. Hinde (Ed.), *Non-verbal communication* (pp. 209–41). Cambridge: Cambridge University Press.

van Schaik, C. P. (2004) *Among orangutans: Red apes and the rise of human culture.* Cambridge, MA: Harvard University Press.

Van Valen, L. (1973) A new evolutionary law. *Evolutionary Theory, 1,* 1–30.

Vandermassen, G. (2005) *Who's afraid of Charles Darwin? Debating feminism and evolutionary theory.* Lanham: Rowman & Littlefield.

von Rueden, C., Gurven, M. & Kaplan, H. (2011). Why do men seek status? Fitness payoffs to dominance and prestige. *Proceedings of the Royal Society B: Biological Sciences, 278,* 2223–32.

Wason, P. C. (1966) Reasoning. In B. Foss (Ed.), *New horizons in psychology.* Harmondsworth, Middlesex, UK: Penguin.

Waynforth, D. (2011) Mate choice and sexual selection. In V. Swami (Ed), *Evolutionary psychology: A critical introduction* (pp. 107–30). Chichester: Wiley-Blackwell.

Waynforth, D., & Dunbar, R. I. M. (1995) Conditional mate choice strategies in humans: Evidence from "lonely hearts" advertisements. *Behaviour, 132,* 755–79.

Weinstein, Y., Bugg, J. M. & Roediger, H. L. (2008) Can the survival recall advantage be explained by basic memory processes? *Memory & Cognition, 36*(5), 913–19.

West, S.A., Griffin, A. S. & Gardner, A. (2007) Social Semantics: Altruism, cooperation, mutualism, strong reciprocity and group selection. *Journal of Evolutionary Biology, 20,* 415–32.

White, T.D., Asfaw, B., Beyene, Y., Haile-Selassie, Y., Lovejoy, C. O., Suwa, G. & WoldeGabriel, G. (2009) *Ardipithecus ramidus* and the paleobiology of early hominids. *Science, 326,* 75–86.

Whiten, A., & Byrne, R. (Eds.) (1988) *Machiavellian intelligence: Social expertise and the evolution of intellect in monkeys, apes and humans.* Oxford: Clarendon Press.

Whiten, A., Goodall, J., McGrew, W. C., Nishida, T., Reynolds, V., Sugiyama, Y., Caroline, E. G., Wrangham, R. W. & Boesch, C. (1999) Cultures in chimpanzees. *Nature, 399*(6737), 682–85.

WHO. (2008) *The global burden of disease: 2004 update.* Geneva: World Health Organization.

WHO. (2012) *WHO depression factsheet, October 2012.* Geneva: World Health Organization.

WHO. (2013) *World health statistics.* Geneva: World Health Organization.

Willer, R., Kuwabara, K. & Macy, M. W. (2009) the false enforcement of unpopular norms. *American Journal of Sociology, 115*(2), 451–90.

Williams, G. C., & Nesse, R. M. (1991) The dawn of Darwinian medicine. *Quarterly Review of Biology, 66*, 1–22.

Wilson, D. S., & Sober, E. (1994) Reintroducing group selection to the human behavioral sciences. *Behavioural and Brain Sciences, 17*, 585–654.

Winegard, B. M., Winegard, B. M. & Deaner, R. O. (2014) Misrepresentations of evolutionary psychology in sex and gender textbooks. *Evolutionary Psychology, 12*, 474–508.

Wood, W., & Eagly, A. H. (2002) A cross-cultural analysis of the behavior of women and men: Implications for the origins of sex differences. *Psychological Bulletin, 128*, 699–727.

Wood, W., & Eagly, A. H. (2012) Biosocial construction of sex differences and similarities in behavior. In J. M. Olson & M. P. Zanna (Eds.), *Advances in experimental social psychology* (Vol. 46, pp. 55–123). London, UK: Elsevier.

Workman, L., & Reader, W. (2014) *Evolutionary psychology* (3rd ed.). Cambridge: Cambridge University Press.

Wrangham, R. (2009) *Catching fire: How cooking made us human.* New York: Basic Books.

Wright, R. (2001) *Nonzero: The logic of human destiny.* New York: Random House.

Wright, R. (2007) *Progress is not a zero-sum game.* TedGlobal [video file] retrieved from https://www.ted.com/talks/robert_wright_on_optimism (May 2014.)

Wynn, K. (1993) Evidence for unlearned numerical competence. *Proceedings of the Joint Annual Meeting of the Western Psychological Association and the Rocky Mountain Psychology Association, Vol. 2.*

Yang, C., Colarelli, S., M., Han, K. & Page, R. (2011) Start-up and hiring practices of immigrant entrepreneurs: An empirical study from an evolutionary psychological perspective. *International Ethnic Entrepreneurship, 20*, 636–45.

Zahavi, A. (1975) Mate selection: A selection for a handicap. *Journal of Theoretical Biology, 53*, 205–14.

Author index

Subject index